HUMAN WORK PRODUCTIVITY

A Global Perspective

HUMAN WORK PRODUCTIVITY

A Global Perspective

Edited by

SHRAWAN KUMAR • ANIL MITAL
ARUNKUMAR PENNATHUR

CRC Press
Taylor & Francis Group
Boca Raton London New York

CRC Press is an imprint of the
Taylor & Francis Group, an **informa** business

CRC Press
Taylor & Francis Group
6000 Broken Sound Parkway NW, Suite 300
Boca Raton, FL 33487-2742

First issued in paperback 2019

ISBN-13: 978-1-4398-7414-1 (hbk)
ISBN-13: 978-0-367-37969-8 (pbk)

Visit the Taylor & Francis Web site at
http://www.taylorandfrancis.com

and the CRC Press Web site at
http://www.crcpress.com

Contents

Preface

Our world is in an economic flux. The relative economic prowess and hence the influence of industry and commerce on the international stage have created a state of unstable and dynamic equilibrium. The interconnectedness of individual productivity, gross domestic product (GDP), global economic impacts, and potential for technological innovation and advancement has never been so apparent. Thus, enhancement of individual productivity exerts a direct impact on personal prosperity, national affluence, and international influence. Based on these connections and impacts, worker productivity has assumed a central and vital place in our individual and collective economic lives.

This is an opportune time to delve into certain aspects of productivity. A topic as broad as productivity deserves and demands an exhaustive discourse—a task that is daunting and beyond the scope of this work. This book is an effort to explore some aspects of productivity in an attempt to expand our understanding of its benefits and consequences.

This book begins with a chapter by Gilad and Elnekave that proposes cost-effective ergonomic improvements to boost individual productivity. These authors measured job elements and determined task durations by extracting elemental times from motion–time tables and analyzing them. Using such a quantitative approach, they were able to improve the operation time and determine appropriate rest allowances. Al Kadeem conversely addresses similar issues by redesign and rearrangement of the workplace to enhance productivity (Chapter 2).

Landau et al. (Chapter 3) discuss the impact of increasing ages of workers in the labor force. Contrary to common belief, they found that declines in productivity due to age were minor. However, they suggest age-related ergonomic adjustments for aging workers to maintain their health.

Alcaraz et al. (Chapter 4) report an investigation on total productive maintenance. They quantified factors using a Likert scale and subjected them to principal component analysis to discern the relative significance of factors affecting productivity.

Production of any plant is impacted directly by maintenance conditions. On the other hand, maintenance activities impede production. To handle this circular problem, Desai and Mital (Chapter 5) investigate the efficacy of workplace design from a maintenance perspective. They found a 58% time saving in maintenance activities resulting from work design modifications. The all-important work hour factor and effects on productivity were studied by Ravindran et al. (Chapter 6). Its significance declined somewhat when the normal work week decreased from 60 hours early in the twentieth century to the current 40 hours. In Chapter 7, the authors compare the productivity of manufacturing workers in the United States with their counterparts in selected developed and emerging economies.

Chapters 8 and 9 address the impacts of technology on productivity. Alcaraz et al. consider the incorporation of advanced manufacturing technologies in an effort to increase productivity. They report a positive outcome of such an approach. In Chapter 9, Minea describes the relationship of technology and productivity as well as techniques for achieving energy savings. Finally, in Chapter 10, Pennathur et al. review factors that impact productivity in healthcare organizations and provide insights on effective organizational designs for improving productivity.

We hope the discussions of several aspects and approaches to productivity issues have addressed at least some of the concerns facing global industries and relevant factors for improving productivity. It is hoped that this small contribution will find a place in the broad economic fields of production and productivity.

Shrawan Kumar
Anil Mital
Arunkumar Pennathur

Editors

Shrawan Kumar, OC, PhD, DSc, FRSC had been a professor of osteopathic manipulative medicine at the University of North Texas Health Science Center in Fort Worth since September 2007. He also served as the director of the Physical Medicine Institute and a graduate advisor for the Physical Medicine Graduate Program at the University of North Texas. Before that, Dr. Kumar was a professor of physical therapy on the Faculty of Rehabilitation Medicine, Division of Neuroscience, at the University of Alberta. He joined the faculty in 1977 and rose to the ranks of associate and full professor in 1979 and 1982, respectively. Currently, he is a professor emeritus of the University of Alberta, Canada.

Dr. Kumar earned a BSc (zoology, botany, and chemistry) and MSc (zoology) from the University of Allahabad, India and a PhD (human biology) from the University of Surrey in the United Kingdom. After completing his PhD program, he did postdoctoral work in engineering at Trinity College, Dublin, Ireland. He also worked as a Council of Scientific and Industrial Research (CSIR) pool officer in orthopedic surgery at the All-India Institute of Medical Sciences, as assistant director of the Central Labour Institute, and as a research associate in the Department of Physical Medicine and Rehabilitation of the University of Toronto, Canada.

Dr. Kumar was recognized for his lifetime work by the University of Surrey and awarded a peer-reviewed DSc in 1994. He was invited to serve as a visiting professor in the Department of Industrial Engineering of the University of Michigan in Ann Arbor in 1983–1984 and was a McCalla Research Professor in 1984–1985.

Dr. Kumar is an Honorary Fellow of the Association of Canadian Ergonomists, a Fellow of the Human Factors and Ergonomics Society of the United States, which awarded him its most prestigious award by naming him a Distinguished International Colleague in 1997. He was also a Fellow of the Ergonomics Society of the United Kingdom, which awarded him the Sir Fredéric Bartlett Medal for excellence in ergonomics research in 1997. In recognition of his distinguished research, he was awarded Killam Annual Professorships for 1997–1998 by the University of Alberta in Canada.

Dr. Kumar was appointed an honorary professor of health sciences at the University of Queensland, Brisbane, Australia in 1998. In 2000, he received the Jack Kraft Innovator Award from the Human Factors and Ergonomics Society of the United States and the Ergonomics Development Award from the International Ergonomics Association for conceptualizing and developing the subdiscipline of rehabilitation ergonomics. In 2002 he received the highest professional honor from the International Ergonomics Association (IEA) by being named a Fellow.

The Ergonomics Society of the United Kingdom invited him to deliver its Annual Society Lecture in 2003. The Royal Society of Canada elected him a Fellow in 2004. The Governor General of Canada named Dr. Kumar an Officer of the Order of Canada in 2009. This is the highest civilian honor awarded in Canada.

Dr. Kumar was honored by King George Medical College, Lucknow, India in 2010 and invited to give a guest lecture to faculty and students. He was named a Distinguished Alumnus by his alma mater, the University of Allahabad, also in 2010. He has been an invited, keynote, or plenary speaker and addressed more than 40 national and international conferences in the United States, Canada, Brazil, the United Kingdom, Germany, Sweden, India, Malaysia, Indonesia, South Africa, Australia, and New Zealand.

Dr. Kumar has written more than 528 scientific peer-reviewed publications and works in the area of musculoskeletal injury causation and prevention with special emphasis on low-back pain and whiplash. He has held grants from National Sciences and Engineering Research Council (NSERC), Medical Research Council (MRC), Workers' Compensation Board (WCB) and National Research Council (NRC). He has supervised the work of 13 MSc students, 7 doctoral candidates, and 6 postdoctoral fellows. He served as an editor of the *International Journal of Industrial Ergonomics*, an assistant editor of *Transactions in Rehabilitation Engineering*, and a consulting editor of *Ergonomics*.

At present, Dr. Kumar is an associate editor of *Spine*, an associate editor of *The Spine Journal*, and an associate editor of *Physical Medicine and Rehabilitation*. He also serves on the editorial boards of eight other international journals and acts as a reviewer for several international peer-reviewed journals and as a grant reviewer for NSERC, Canadian Institute of Health Research (CIHR), Alberta Heritage Foundation of Medical Reseach (AHFMR), Alberta Occupational Health and Safety, British Columbia Research, the Wellcome Foundation in England, and the National Institutes of Health and the National Science Foundation in the United States.

Dr. Kumar organized and chaired two highly successful international conferences along with a national meeting and several regional conferences. He served as secretary and president of the International Society of Occupational Ergonomics and Safety, chair of the Graduate Program in Physical Therapy from 1979–1987, director of research from 1985–1990, chair of the Doctoral Program Development Committee of the Faculty of Rehabilitation Medicine, and as a member of several other committees. Dr. Kumar was a member of the University of Alberta Planning and Priority Committee and Academic Development Committee. He formerly chaired the Code of Ethics Subcommittee of the International Ergonomics Association, served as the president of the Indo-Canadian Society and Council of India Societies of Edmonton, and is currently a patron of the Society for Development in Third World Countries.

Anil Mital, BE, MS, PhD is a professor of mechanical engineering and manufacturing engineering and design and also a professor of physical medicine and rehabilitation at the University of Cincinnati. Earlier, he was a professor and director of industrial engineering at the same institution.

Dr. Mital earned a BE in mechanical engineering from Allahabad University, India, and an MS and PhD in industrial engineering from Kansas State University and Texas Technical University, respectively.

Dr. Mital is the founding editor-in-chief emeritus of both the *International Journal of Industrial Ergonomics* and the *International Journal of Industrial Engineering*. He is also the former executive editor of the *International Journal of Human Resource Management and Development* and author or editor of more than 500 technical publications including 24 books. Dr. Mital's current research interests include design and analysis of human-centered manufacturing systems, application of DFX principles to product design, economic justification, and manufacturing planning and facilities design.

Dr. Mital founded the International Society of Occupational Ergonomics and Safety and received its first Distinguished Accomplishment Award (1993). He is a Fellow of the Human Factors and Ergonomics Society and a recipient of its Paul M. Fitts Education Award (1996) and Jack A. Kraft Innovator Award (2012).

Dr. Mital is also a recipient of the Liberty Mutual Insurance Company's Best Paper Award (1994). He was named a Fellow of the Institute of Industrial Engineers and is a past director of its Ergonomics Division. In 2007, Dr. Mital received the Dr. David F. Baker award from the Institute of Industrial Engineers for lifetime research activities. He has also received the Eugene L. Grant Award from the American Society of Engineering Education and the Ralph R. Teetor Award from the Society of Automotive Engineers.

Arunkumar Pennathur, PhD focuses his research expertise and experience on human factors engineering, including physical and mental workload modeling projects for the U.S. Army. His research in modeling cognitive work in planning has been funded by the National Science

Foundation (NSF). His human factors work on modeling the etiology and progression of disability and disease among older adults and developing gerontechnology interfaces and engineering interventions for promoting graceful and successful aging with usable living spaces was funded by the National Institutes of Health (NIH).

Dr. Pennathur writes about and teaches sociotechnical approaches to work design. He has accumulated very significant qualitative analysis and modeling experience and conducted emergent themes analysis with qualitative text and video content analysis. He also has performed quantitative data analysis in all his research work.

Dr. Pennathur has published more than 100 peer-reviewed papers, book chapters, and books and presented his work at numerous national and international conferences on human factors engineering and industrial engineering. He has developed undergraduate and graduate curricula in human factors engineering and trained students in advanced ergonomics, advanced safety engineering, advanced work design, cognitive work analysis, and design and analysis of experiments, among other topics.

Dr. Pennathur is the editor-in-chief of the *International Journal of Industrial Engineering* and the founding editor-in-chief of the *Journal of Applications and Practices in Engineering Education*. He serves on the editorial board of the *Open Journal on Ergonomics* and is a member of the Human Factors and Ergonomics Society.

Contributors

Jorge Luis García Alcaraz
Institute of Engineering and
  Technology
Autonomous University of
  Ciudad Juárez
Ciudad Juárez, Chihuahua,
  México

Reem Al Kadeem
Faculty of Engineering
Alexandria University
Alexandria, Egypt

Regina Brauchler
Ergonomia GmbH & Co.
Stuttgart, Germany

Paulina Cano
Sands Research
El Paso, Texas

Anoop Desai
Department of Mechanical and
  Electrical Engineering
Georgia Southern University
Statesboro, Georgia

Moran Elnekave
Microsoft Corporation
Redmond, Washington

Natalia Espino
Industrial Engineering
  Department
University of Texas at El Paso
El Paso, Texas

Issachar Gilad
Faculty of Industrial Engineering
  and Management
Technion–Israel Institute of
  Technology
Haifa, Israel

Alejandro Alvarado Iniesta
Institute of Engineering and
  Technology
Autonomous University of
  Ciudad Juárez
Ciudad Juárez, Chihuahua,
  México

Kurt Landau
Institut für Organisation und
  Arbeitsgestaltung GmbH
Millstatt, Austria

Jacqueline Loweree
Anthropology Department
University of Texas at El Paso
El Paso, Texas

Aidé Aracely Maldonado Macías
Institute of Engineering and
  Technology
Autonomous University of
  Ciudad Juárez
Ciudad Juárez, Chihuahua,
  México

Sergio Gutiérrez Martínez
Institute of Engineering and
  Technology
Autonomous University of
  Ciudad Juárez
Ciudad Juárez, Chihuahua,
  México

Alina Adriana Minea
Faculty of Materials Science and
  Engineering
Universitatea Tehnică Gheorghe
  Asachi
Iaşi, Romania

Aashi Mital
Department of History
University of Cincinnati
Cincinnati, Ohio

Anil Mital
Department of Mechanical
  Engineering
University of Cincinnati
Cincinnati, Ohio

Salvador Noriega Morales
Institute of Engineering and
  Technology
Autonomous University of
  Ciudad Juárez
Ciudad Juárez, Chihuahua,
  México

Arunkumar Pennathur
Mechanical and Industrial
  Engineering
University of Iowa
Ames, Iowa

Priyadarshini Pennathur
Department of Mechanical and
  Industrial Engineering
University of Iowa
Ames, Iowa

Angelika Presl
Klinik Bavaria
Kreischa, Germany

Vignesh Ravindran
Department of Mechanical
  Engineering
University of Cincinnati
Cincinnati, Ohio

Margit Weißert-Horn
Ergonomia GmbH & Co.
Stuttgart, Germany

chapter one

Ergonomics improvements for the human operator— Cost effectiveness approach

Issachar Gilad and Moran Elnekave

Contents

1.1 Introduction

The cost effectiveness of ergonomic solutions is often questioned by manufacturing managers and administrative functions in organizations. Practice shows again and again that managers omit ergonomic solutions because of their high cost. The aim of a study conducted at Technion–Israel Institute of Technology was to develop a practical approach to implement cost effective ergonomic solutions by coupling a computerized job description design and predetermined motion–time systems. This is done by generating a comparison of operation times and body motions used in existing and improved (safer) work situations.

This chapter introduces a quantified approach and demonstrates the implementation of ergonomic solutions in an industrial case study. The authors explain in the case study how a solution may impact operation times and worker stress and how it can be beneficial in terms of relaxation allowances required for resting and recovery. The final results— ergonomic recommendations of interest to managers—are presented in easy-to-understand figures and numbers. Managers can then easily calculate the costs and potential return on investment for an ergonomic solution.

An unsuitable match of human body capabilities and job require-ments produces palpable consequences. Physical limitations caused by pain and stress to which the human body is subjected cause loss of effi-ciency and can even lead to disability. The failure to match work hazards or repetitive work requirements with human capabilities always results in reduced production.

Awareness of ergonomics has grown during recent decades and risk factors are now measured and analyzed from quantitative data instead of hunches. Work measurement and ergonomic analysis enable the observa-tion of worker activities and the evaluation of work efficiency based on time metrics and body part utilization. The more quantitative the data obtained, the greater the pragmatic importance and the stronger the reflection of the reality of a given operation. Relevant and accurate data answer the needs of practitioners seeking effective solutions for solving industrial worker problems created by repetitious activities and improper combinations of humans and machines.

Engineers responsible for designing workplaces and tools are becom-ing less conservative and no longer accept existing designs as inevitable. Consequently, they make more effort early in the design states to develop work situations that match operator capabilities and take into consider-ation the ability of an operator to adapt to localized stress. To meet opera-tor needs effectively, it is important to evaluate tasks quantitatively. The scientific procedure that deals with quantitative task evaluation is the time and motion study—also known as a work study and a partner of ergonomics in designing better work conditions. Ergonomics and work study techniques are similar in the following areas:

- Both techniques are used to maximize work efficiency. Ergonomics is defined as the research of factors influencing human work effi-ciency. Conversely, efficiency is maximized when a standard work method exists and performance time standards are defined and derived by implementing scientific approaches—work measure-ment techniques. Coupling ergonomics and work study should lead to development of a system that improves efficiency based on mini-mal operation time, effort, and cost.
- Both techniques rely on the basics of motion study and motion econ-omy. Although developed empirically, these principles are based on direct correlation of human anatomical, biomechanical, and physi-ological principles to a specific task.

In this sense, the basics of both disciplines merge because they have a common aim or improving work efficiency by designing and implementing optimal motions in work operations.

The proposed approach is unique because it involves a complete design by computing results for existing work situations and forecasting results for improvements based on cost comparisons. Typically, many ergonomic practitioners find it difficult to visualize the effects of workstation changes on worker comfort and operation times. To help them grasp the implications of a redesign, a detailed mock-up phase that should include time study and ergonomic analysis is often devised.

Modern computer-aided design (CAD) tools are available to assist designers to analyze work situation safety. Nevertheless, these tools are expensive and require many hours for modeling basic work situations (Ben-Gal and Bukchin 2002). When complexity of motion is included in a work situation to be studied, the cost effectiveness of such tools is questionable. Moreover, when using software that supports human motion, operation times act as inputs and thus a predetermined motion time system (PMTS) study should be performed to predict times accurately. The authors of this chapter described a generalized process for using such immersive tools (Elnekave and Gilad 2006).

1.2   Methodology

After hazardous work elements are identified in the work cycle using the suggested quantitative method, ergonomic-driven improvements are suggested. After that, the costs of ergonomic improvements can be formulated for various redesign alternatives. The alternatives allow management to make decisions on the basis of investment and expected return through savings derived from implementing ergonomic solutions. The stages of the methodology are outlined in Figure 1.1 and explained below. The procedure for identifying hazardous work elements was reviewed in Gilad and Elnekave 2006a. Once hazardous work elements are identified, they should be targeted for intervention and improvement, as demonstrated in Gilad and Elnekave 2006b.

1.2.1   Design improvements

Ergonomic improvements must focus on preventing and/or relieving the identified hazards. When considering designs and improvements for man–machine work situations, human limitations should be treated as constraints that will help drive the ergonomic improvement. The following table lists improvements based on identified hazards:

Body Area	Hazard
Back	Continuous static effort on the lower back will ultimately cause pain. Such pain may be minimized if a worker keeps an erect back posture. Leaning and bending elements should be improved or eliminated by adjusting normal working heights.
Shoulder	Continuous static effort can cause shoulder joint arthritis. Keeping hands close to the body minimizes the impact. Obstacles constraining normal shoulder posture should be changed.
Elbow	Continuous static effort may disrupt normal elbow functioning. Examples are holding objects too long and exerting forces with upper extremities while keeping elbows straight. The impacts may be avoided by changing work area distances to allow workers to bend their elbows.
Wrist	Continuous static effort can lead to disability of hand movements. Extended periods of ulnar deviation during a work cycle and repetitive snapping of the wrist in flexion and extension motions cause this. Tool use can lead to poor posture. Maintaining correct working heights or purchasing tools that support good posture may improve such flaws. Correction is not easy when the same tool is used for several tasks at different working heights.
Neck	Continuous static effort will eventually cause pain in muscle groups around the neck. Keeping the neck in an upright posture will prevent fatigue to the muscles supporting it and avoid cumulative neck trauma disorders.

Common inexpensive tools can be used for visualizing work situations. Visio has been used as a sketching board and is commonly used by industrial engineers. To determine a correct working posture, a mannequin is positioned and its joint motions manipulated to reflect a common posture for the studied work element. Then the heights of the work surfaces are manipulated to impose correct posture on upper body joint motions. One feature of many workstation improvements is maintaining joints in neutral areas to eliminate poor posture. It is important to analyze joint ranges and motions in sagittal, transverse, and planar views.

When geometric constraints do not allow a match between a human operator and a workstation, mechanical assistance is an alternative solution but such measures vary in performance, cost, and impact on body posture. It is always best to select a tool that facilitates a correct working posture, especially if a more ergonomic tool costs about the same and performs as well as other options. However, if an ergonomic solution reveals

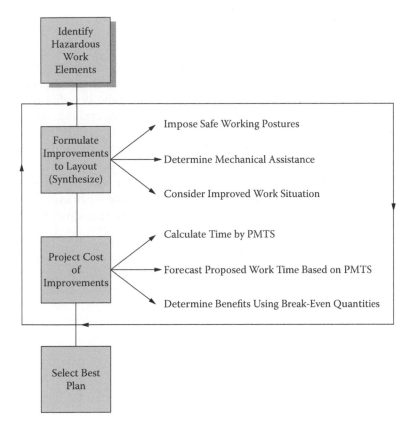

Figure 1.1 Cost effectiveness in ergonomic design.

big differences in price and performance, the proposed improvements must be evaluated further.

The duration of static posture required by a worker based on a motion and time study must be known before costly improvements in work situations are considered. The analyst must quantitatively reflect changes in a work situation if solid economic evaluation of a redesign is required. For example, working height is at the root of poor posture caused by lifting of bulky objects. Ideally, objects should be kept at safe working heights (Saleem et al. 2003). In many cases, material is stacked and bending heights change as the work cycle advances. A common approach is to purchase a lift for containers to reduce bending to grasp objects. Another approach is to set raw materials to be piled vertical to the floor to allow grasping without bending. In both cases, the analyst must project the savings in time and effort in comparison to the investment made.

1.2.2   Project improvement costs

In order to evaluate the cost of improvements, a predetermined motion–time study (PMTS) of the existing and proposed work situations must be made. The general approach is to compare the existing and proposed work situation costs quantitatively and determine the benefits in making improvements by analyzing the number of work cycles required to achieve return on investment.

Equation (1.1) calculates the number of work cycles needed to return an invested cost for an ergonomic improvement. The invested costs essential to implement an ergonomic improvement are divided by the savings that result from implementing the improvement in the work cycle to yield the number of cycles required to "break even" on the investment.

$$BEQ_i \frac{\text{Investment in Improvement } i}{\text{Savings from Improvement } i \text{ per work cycle}} = \frac{IC_i}{\Delta SCT \cdot LC} \quad (1.1)$$

$$\Delta SCT = SCT_{existing} - SCT_{proposed} \quad (1.2)$$

$$SCT = \sum_i [NOT_j \cdot (1 - RA_j)] \quad (1.3)$$

where:

    BEQ_i = break-even quantity of cycles returning investment on improvement i.
    IC_i = invested costs in implementing ergonomic improvement i.
    ΔSCT = difference between existing and proposed standard cycle times.
    LC = labor costs.
    SCT = standard cycle time of existing and proposed work situations.
    NOT_j = normal operation time of work element j from PMTS study.
    RA_j = resting allowances for work elements j.

Savings from implementing improvements are calculated as the difference between standard cycle times of existing and proposed work cycles multiplied by the labor costs as shown in Equation (1.1). Equation (1.2) describes the difference in cycle time between the existing and proposed work situations. Equation (1.3) shows the cycle time calculation. Normal times of work elements are multiplied by their corresponding resting allowances. Resting allowance tables are the most practical options for converting ergonomic factors into time metrics. Normal operation times for existing and proposed work situations are calculated using the PMTS.

1.2.3   Selecting the best plan

The best plan is not unequivocal. It depends on managerial strategy for investing in improvement. Some improvements return investments quite quickly and are favored as a result. Others may be difficult for management to accept because they may lengthen operation times and thus slow production. Therefore, each ergonomic improvement should be communicated to management separately and no improvement should be considered a full solution.

Improvements should be ranked according to the rate of return on investment to help management decide its options. In many cases, as sad as it sounds, management may decide not to pursue ergonomic improvements and continue production without increasing cost. The implications of such decisions are explained in Section 1.4.

1.3   Case study

A proposed approach for redesign was considered for an industrial case study involving assembly of a bed linen box. The sub-operation normal time and operation sequence were calculated using BasicMOST (Maynard operation sequence technique). Times are shown as time measurement units (TMUs) and 1 TMU = 0.0006 min (Zandin 1990).

Allowances were calculated using standard ILO (International Labor Office) tables for each work element. Table 1.1 shows a BasicMOST analysis for the existing work situation for assembly of a bed linen box. A normal time for performing the work cycle was 5.71 min (9520 TMUs). Allowances were calculated at 11.8%, leading to a standard time of 6.39 min (10,647 TMUs), leaving 0.68 min of rest per work cycle.

Screw fastening operations were found to be most hazardous. The observed poor posture for this work element in the filmed work situation arose from the need to operate a power tool at obstructed points in the assembly; the corners were difficult to reach because of the box geometry. Figure 1.2 compares existing and improved postures. Improvement 1 suggested changing the operation sequence by first screwing the corner support to the long boards (laid flat on the workbench) before nailing them to the short boards. This way the points for attaching screws are not obstructed. To foster a correct working posture, the current pistol-shaped screwdriver must be replaced by a vertical one.

Table 1.2 formulates the proposed Improvement 1 in BasicMOST. Changes in the operation sequence are highlighted in the proposed work sequence. Changing the operation sequence imposes more tool changing operations and actually lengthens the operative work cycle. The better posture improves allowances for positioning the corner supports and screwing them onto the long boards. In addition, motions are

Table 1.1 BasicMOST Analysis of Existing Work Situation in Case Study of Bed Linen Box Assembly

Operation		BasicMOST Analysis											Sub-Operation Frequency	Normal Operation Time (TMUs)	Allowances (%)	Standard Operation Time (TMUs)
Load 2 long boards	Sub-Operation Sequence	A_{10}	B_3	G_3	A_{10}	B_0	P_1	A_1					1	280	6	297
	Partial Frequency	1	1	1	1	]										
Load 2 short boards	Sub-Operation Sequence	A_{10}	B_3	G_3	A_{10}	B_0	P_1	A_0					1	270	6	286
	Partial Frequency	1	1	1	1	]										
Position long board to short board	Sub-Operation Sequence	A_6	B_0	G_1	A_1	B_0	P_0	A_0					4	1,040	10	1,144
	Partial Frequency	1	2	2	1											
Attach long board to short board with three nails using nail gun	Sub-Operation Sequence	A	B_0	G_1	A_1	B_0	P_3	F_1	A	B_0	P_1	A_0	4	920	14	1,049
	Partial Frequency	1	1	3	3	3	1	1	]							
Load 2 rods	Sub-Operation Sequence	A_{10}	B_3	G_1	A_{10}	B_0	P_1	A_0					1	250	4	260
	Partial Frequency	1	1	1	1											
Position rod to long board	Sub-Operation Sequence	A_{10}	B_0	G_1	A_3	B_0	P_3	A_0					4	680	5	714
	Partial Frequency	1	1	1	1											
Attach rod to long board with nail using nail gun	Sub-Operation Sequence	A_1	B_0	G_1	A_1	B_0	P_3	F_1	A_1	B_0	P_1	A_0	4	360	13	407
	Partial Frequency	1	1	1	1	1	1	1	]							
Change tool from nail gun to screwdriver	Sub-Operation Sequence	A_1	B_0	G_1	M_1	L_0	A_0						1	30	6	32
	Partial Frequency	1	1	1												
	Sub-Operation Sequence	A_0	B_0	G_0	A_1	B_0	P_1	A_D					1	20	6	21

Operation	Row	Sequence / Partial frequencies				Total
	Partial Frequency	1 1 1 1 1 1 1				
	Sub-Operation Sequence	A₁ B₀ G₁ A₁ B₀ P₃ A_D	1	60	6	64
Position corner support	Partial Frequency	1 1 2 1 1 1 1				
	Sub-Operation Sequence	A₃ B₀ G₁ A₃ B₀ P₀ A₀	4	560	7	599
Assemble corner support with 4 screws using power screwdriver	Partial Frequency	1 1 1 4 4 4 1 1 1 1 1				
	Sub-Operation Sequence	A₁ B₀ G₁ A₁ B₀ P₁ F₃ A₁ B₀ P₁ A₀	4	1,280	13	1,446
Change tool from screwdriver to nail gun	Partial Frequency	1 1 1 1 1 1 1				
	Sub-Operation Sequence	A₁ B₀ G₁ M₁ X₀ I₀ A₀	1	3D	6	32
	Partial Frequency	1 1 1 1 1 1 1				
	Sub-Operation Sequence	A₀ B₀ G₀ A₁ B₀ P₁ A₀	1	2D	6	2]
	Partial Frequency	1 1 1 1 1 1 1				
	Sub-Operation Sequence	A₁ B₀ G₁ A₁ B₀ P₁ A₀	1	6D	6	64
Load backboard	Partial Frequency	1 1 1 1 1 1 1				
	Sub-Operation Sequence	A₁₀ B₃ G₃ A₁₀ B₀ P₁ A₀	1	29D	7	310
Position backboard	Partial Frequency	1 1 1 1 1 1 1				
	Sub-Operation Sequence	A₃ B₀ G₃ A₁ B₀ P₆ A₀	1	13D	10	143
	Partial Frequency	1 1 1 1 1 1 1				

(continued)

Table 1.1 BasicMOST Analysis of Existing Work Situation in Case Study of Bed Linen Box Assembly (continued)

	BasicMOST Analysis											Sub-Operation Frequency	Normal Operation Time (TMUs)	Allowances (%)	Standard Operation Time (TMUs)
Attach backboard to boards and rods with 54 nails using nail gun															
Sub-Operation Sequence	A_3	B_0	G_1	A_1	B_0	P_3	F_1	A_1	B_0	P_1	A_1	1	2,770	17	3,241
Partial Frequency	1	1	1	54	54	54	54	1	1	1	1				
Unload backboard															
Sub-Operation Sequence	A_3	B_3	G_3	A_{16}	B_3	P_3	A_{16}					1	470	10	517
Partial Frequency	1	1	1	1	1	1	1								

Normal cycle time: 9,520 TMUs

Allowance per cycle: 11.8%

Standard cycle time: 10,647 TMUs

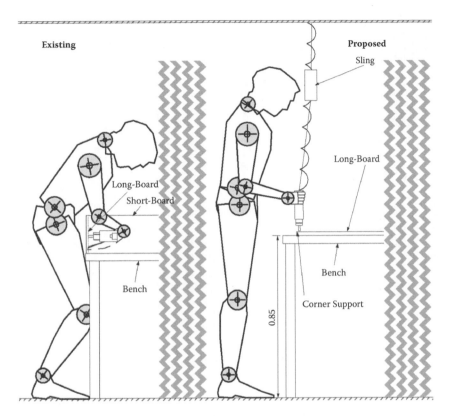

Figure 1.2 Postures of existing and proposed designs for screwing corner supports.

shortened for positioning the corner supports and the screwdriver. As a result, positioning locations are easy to reach and not obstructed.

A section at the bottom of the table summarizes the comparison of the existing and improved work situation. The proposed work situation improves the standard time by 25 TMUs per work cycle even though the earlier work cycle is lengthened by 20 TMUs. The total saving per workday is only $0.37. Assuming a $20 per hour cost for the worker and an investment of $250 for a vertical screwdriver, the investment will be returned within 669 workdays.

Improvement 2 suggested using a sling to reduce the worker force required to overcome the weight of the power screwdriver (Figure 1.3). Such an improvement would reduce the allowances for the screwing operation by 2% without affecting operation times. The result will be a 25.6-TMU improvement to the standard cycle time and $0.39 saving per workday. Assuming an approximate cost of $150 for a sling, this improvement will be returned in 389 days. This improvement was not formulated

Table 1.2 BasicMOST Analysis of Proposed Work Situation Integrating Improvement 1[a]

Operation		BasicMOST Analysis											Sub-Operation Frequency	Normal Operation Time (TMUs)	Allowances (%)	Standard Operation Time (TMUs)
Load 2 long boards	Sub-Operation Sequence	A_{10}	B_3	G_3	A_{10}	B_0	P_1	A_1					1	280	6	297
	Partial Frequency	1	1	1	1	1	1	1								
Position corner support	Sub-Operation Sequence	A_3	B_0	G_1	A_3	B_0	P_3	A_0					4	440	5	462
	Partial Frequency	1	1	1	1	1	1									
Change tool from nail gun to screwdriver	Sub-Operation Sequence	A_1	B_0	G_1	M_1	X_0	I_0	A_0					1	30	6	32
	Partial Frequency	1	1	1	1											
	Sub-Operation Sequence	A_0	B_0	G_0	A_1	B_0	P_1	A_0					1	20	6	21
	Partial Frequency	1	1	1	1											
	Sub-Operation Sequence	A_1	B_0	G_1	A_1	B_0	P_3	A_0					1	60	6	64
	Partial Frequency	1	1	1	1											
Attach corner support to long board with 3 screws using power screwdriver	Sub-Operation Sequence	A_1	B_0	G_1	A_1	B_0	P_1	F_3	A_1	B_0	P_1	A_0	4	760	9	828
	Partial Frequency	1	1	3	3	1	1	1	1							
Load 2 short boards	Sub-Operation Sequence	A_{10}	B_3	G_3	A_{10}	B_0	P_1	A_0					1	270	6	286
	Partial Frequency	1	1	1	1	1	1									
Change tool from screwdriver to nail gun	Sub-Operation Sequence	A_1	B_0	G_1	M_1	X_0	I_0	A_0					1	30	6	32
	Partial Frequency	1	1	1	1											
	Sub-Operation Sequence	A_0	B_0	G_0	A_1	B_0	P_1	A_0					1	20	6	21
	Partial Frequency	1	1	1	1											

Operation		1	2	3	4	5	6	7	8	9	10	11				
	Partial Frequency	1	1	1	1	1	1	1								
	Sub-Operation Sequence	A_1	B_0	G_1	A_1	B_0	P_3	A_0					1	60	6	64
Position long board to short board	Partial Frequency	1	1	2	1	1	1	1								
	Sub-Operation Sequence	A_0	B_0	G_1	A_1	B_0	P_0	A_0					4	1,040	10	1,144
Attach long board to short board with 3 nails using nail gun	Partial Frequency	1	2	2	1	1	2	1	1	1	1	1				
	Sub-Operation Sequence	A_3	B_0	G_1	A_1	B_0	P_3	F_1	A_3	B_0	P_1	A_0	4	920	14	1,049
Load 2 rods	Partial Frequency	1	1	1	3	3	3	3								
	Sub-Operation Sequence	A_{10}	B	G_1	A_{10}	B_0	P_1	A_0					1	250	4	260
Position rod to long board	Partial Frequency	1	1	1	1	1	1	1								
	Sub-Operation Sequence	A_{10}	B_0	G_1	A_3	B_0	P_3	A_0					4	680	5	714
Attach rod to long board with nail gun	Partial Frequency	1	1	1	1	1	1	1	1	1	1	1				
	Sub-Operation Sequence	A_1	B_0	G_1	A_1	B_0	P_3	F_3	A_1	B_0	P_1	A_0	4	360	13	407
Change tool from nail gun to screwdriver	Partial Frequency	1	1	1	1	1	1	1								
	Sub-Operation Sequence	A_1	B_0	G_1	M_1	X_0	L_0	A_0					1	30	6	32
	Partial Frequency	1	1	1	1	1	1	1								
	Sub-Operation Sequence	A_0	B_0	G_0	A_1	B_0	P_1	A_0					1	20	6	21
	Partial Frequency	1	1	1	1	1	1	1								
	Sub-Operation Sequence	A_1	B_0	G_1	A_1	B_0	P_3	A_0					1	60	6	64
	Partial Frequency	1	1	1	1	1	1	1								

(continued)

Table 1.2 BasicMOST Analysis of Proposed Work Situation Integrating Improvement 1[a] (continued)

Operation		BasicMOST Analysis	Sub-Operation Frequency	Normal Operation Time (TMUs)	Allowances (%)	Standard Operation Time (TMUs)
Attach corner support with screw to short board using power screwdriver	Sub-Operation Sequence	A_1 B_0 G_1 A_1 B_0 P_3 F_3 A_1 B_0 P_1 A_0	4	440	13	497
	Partial Frequency	1 1 1 1 1 1 1 1 1 1				
Change tool from screwdriver to nail gun	Sub-Operation Sequence	A_1 B_0 G_1 M_1 X_0 I_0 A_0	1	30	6	32
	Partial Frequency	1 1 1 1 1				
	Sub-Operation Sequence	A_0 B_0 G_0 A_1 B_0 P_1 A_0	1	20	6	21
	Partial Frequency	1 1 1				
	Sub-Operation Sequence	A_1 B_0 G_1 A_1 B_0 P_3 A_0	1	60	6	64
	Partial Frequency	1 1 1				
Load backboard	Sub-Operation Sequence	A_{10} B_3 G_3 A_{10} B_0 P_3 A_0	1	290	7	310
	Partial Frequency	1 1 1				
Position backboard	Sub-Operation Sequence	A_3 B_0 G_3 A_1 B_0 P_6 A_0	1	130	10	143
	Partial Frequency	1 1 1				
Attach backboard to boards and rods with 54 nails using nail gun	Sub-Operation Sequence	A_3 B_0 G_1 A_1 B_0 P_3 F_1 A_1 B_0 P_1 A_1	1	2,770	17	3,241
	Partial Frequency	1 1 54 54 54 1 1 1				

Unload backboard	Sub-Operation Sequence											
Partial Frequency	A_3	B_3	G_3	A_{10}	B_3	P_1	A_{16}	1	470	10	517
	1	1	1	1	1	1	1				

Normal cycle time: 9,540 TMUs

Allowance per cycle: 11.3%

Standard cycle time: 10,622 TMUs

Calculations of savings and break-even point

Difference between existing and proposed standard cycle times:	25 TMUs/work cycle
Difference between existing and proposed normal cycle times:	20 TMUs/work cycle
Hourly labor cost:	$20
Operational savings for 8-hour workday:	$0.30
Savings in allowances for 8-hour workday:	$0.67
Total savings for 8-hour workday:	**$0.37**
Invested Cost:	$250
BEQ:	**669 workdays**

[a] Improvement 1: Changing operation sequence (attaching corner support to long board before attaching long board to short board) and using vertical screwdriver during corner assembly to impose safe working posture.

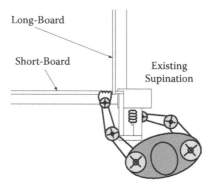

Existing posture for nailing long-board to short-board

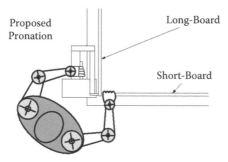

Proposed posture for nailing long-board to short-board

Figure 1.3 Postures of existing and proposed designs for nailing operations.

in a separate table because it does not impose any change in the operation sequence; it simply alleviates worker stress and is expressed as allowances.

The second most hazardous work element is the long- to short-board nailing operation performed using a power nail gun. This work sets up the initial box shape. In the existing situation, the task is carried out in a problematic 90-degree supination (Figure 1.4). Due to the geometry of the assembly, the operation cannot be improved easily, as was the case for the situation described above, without adding many motions to the sequence. Nevertheless, the worker on this operation is better off positioning the nail gun with a 90-degree pronation rather than supination (Figure 1.4).

This change is designated Improvement 3. It will not require a capital investment. The plant's engineering team simply must re-educate the worker. Improvement 3 reduces allowances for these nailing operations by 1% due to better posture and produces no changes in operation times. The measure will save 9.2 TMUs per work cycle (about $0.14 per work-

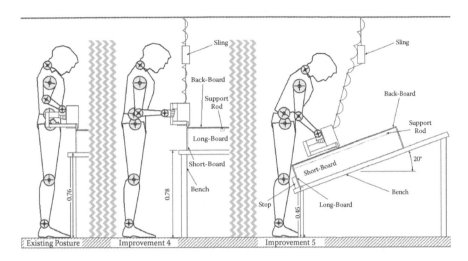

Figure 1.4 Postures of existing and proposed designs for backboard nailing.

day). As no expenditure is required, such an improvement is favorable but relies on workers for implementation.

Improvements were considered for the backboard nailing operations as well. Improvement 4 suggested replacing a pistol-shaped nail gun with a vertical type that would allow a worker to maintain normal posture. Improvement 5 proposed keeping the pistol-shaped nail gun and inclining the work surface by 20 degrees to allow correct working posture. In the backboard nailing backboard operation, a worker uses a pistol tool directed perpendicular to the ground and this causes ulnar deviation. Moreover, since the backboard is located roughly at elbow height and does not take into consideration the tool, shoulder adduction is also inevitable. Figure 1.4 presents the existing work posture and Improvements 4 and 5 for imposing correct work posture during backboard nailing.

Improvement 4 involves only working posture and does not influence the operation sequence. The improvement will reduce allowances by 2% for backboard nailing operations, resulting in a saving of 55.4 TMUs per work cycle ($0.83 per workday based on an hourly labor cost of $20). Replacing the pistol-shaped tool with a vertical tool will likely cost about $250 that may be returned in about 299 workdays.

Table 1.3 formulates the BasicMOST sequence for Improvement 5, which includes reorientation of the box because the inclination will impose poor posture when a worker attempts to work on the high side of the assembly (about 35 cm higher). Reorientation operations are highlighted in the table, along with extra operations for grasping and releasing tools at hand during backboard nailing and screwing operations. Improvement 5 will reduce stress, expressed as a 2% reduction of

Table 1.3 BasicMOST Analysis of Proposed Work Situation Integrating Improvement 5[a]

Operation		BasicMOST Analysis	Sub-Operation Frequency	Normal Operations Time (TMUs)	Allowances (%)	Standard Operation Time (TMUs)
Load 2 long boards	Sub-Operation Sequence	A_{10} B_3 G_3 A_{10} B_0 P_1 A_1	1	280	6	297
	Partial Frequency	1 1 1 1 1 1				
Load 2 short boards	Sub-Operation Sequence	A_{10} B_3 G_3 A_{10} B_0 P_1 A_1	1	270	6	286
	Partial Frequency	1 1 1 1 1 1				
Position long board to short board	Sub-Operation Sequence	A_6 B_0 G_3 A_2 B_0 P_6 A_0	4	1,040	10	1,144
	Partial Frequency	1 1 2 1 2 1				
Attach long board to short board with 3 nails using nail gun	Sub-Operation Sequence	A_3 B_0 G_1 A_1 B_0 P_3 F_1 A_3 B_0 P_1 A_0	4	920	14	1,049
	Partial Frequency	1 3 3 3 1 1 1				
Load 2 rods	Sub-Operation Sequence	A_{10} B_3 G_1 A_{10} B_0 P_1 A_0	1	250	4	260
	Partial Frequency	1 1 1 1 1				
Position rod to long board	Sub-Operation Sequence	A_{10} B_0 G_1 A_3 B_0 P_3 A_0	4	680	5	714
	Partial Frequency	1 1 1 1 1				
Attach rod to long board with nail using nail gun	Sub-Operation Sequence	A_1 B_0 G_1 A_1 B_0 P_3 F_1 A_3 B_0 P_1 A_0	4	360	13	407
	Partial Frequency	1 1 1 1 1 1				
Change tool from nail gun to screwdriver	Sub-Operation Sequence	A_1 B_0 G_1 M_1 X_0 I_3 P_3 A_1	1	30	6	32
	Partial Frequency	1 1 1 1 1				
	Sub-Operation Sequence	A_1 B_0 G_0 A_1 B_0 P_3 A_0				

Operation																
	Sub-Operation Sequence	A_1	B_0	G_1	A_2	B_0	P_3	A_0								
	Partial Frequency	1	1	1	1	1	1	1					1	20	6	21
Position corner support	Sub-Operation Sequence	A_3	B_0	G_1	A_3	B_0	P_6	A_0								
	Partial Frequency	1	1	2	1	1	1	1					1	60	6	64
Assemble corner support with 4 screws using power screwdriver	Sub-Operation Sequence	A_1	B_0	G_1	A_1	B_0	P_3	F_3								
	Partial Frequency	1	1	2	1	1	1	1					4	560	7	599
Reorient assembly	Sub-Operation Sequence	A_1	B_2	G_2	A_1	B_3	P_3	A_0	A_1	B_0	P_1	A_0				
	Partial Frequency	2	1	2	4	4	4	4	2	1	2	1	4	1,440	13	1,627
	Sub-Operation Sequence	A_1	B_2	G_2	A_1	B_3	P_3	A_0								
	Partial Frequency	1	1	1	2	1	1	3					1	150	10	165
Change tool from screwdriver to nail gun	Sub-Operation Sequence	A_1	B_0	G_1	M_1	X_3	I_3	A_0								
	Partial Frequency	1	1	1	1	1	1	1					1	30	6	32
	Sub-Operation Sequence	A_0	B_0	G_1	A_1	B_3	P_3	A_3								
	Partial Frequency	1	1	1	1	1	1	1					1	20	6	21
Load backboard	Sub-Operation Sequence	A_{10}	B_3	G_3	A_{10}	B_0	P_3	A_3								
	Partial Frequency	1	1	1	1	1	1	1					1	60	6	64
Position backboard	Sub-Operation Sequence	A_{10}	B_3	G_3	A_2	B_0	P_3	A_3								
	Partial Frequency	1	1	1	1	1	1	1					1	290	7	310
	Sub-Operation Sequence	A_3	B_0	G_3	A_2	B_0	P_6	A_0								
	Partial Frequency	1	1	1	1	1	1	1					1	130	10	143

(continued)

Table 1.3 BasicMOST Analysis of Proposed Work Situation Integrating Improvement 5[a] (continued)

		BasicMOST Analysis												Sub-Operation Frequency	Normal Operations Time (TMUs)	Allowances (%)	Standard Operation Time (TMUs)
Attach backboard and rods with 54 nails using nail gun	Sub-Operation Sequence	A_1	B_0	G_1	A_1	B_0	P_1	A_1	F_1	A_1	B_0	P_1	A_1	1	2,830	15	3,255
	Partial Frequency	2	1	2	54	54	54	54	54	2	1	1	2				
Reorient assembly	Sub-Operation Sequence	A_2	B_2	G_2	A_1	B_3	P_3	A_0						1	150	10	165
	Partial Frequency	1	1	1	2	1	1	1									
Unload backboard	Sub-Operation Sequence	A_3	B_3	G_3	A_{10}	B_3	P_3	A_{16}						1	470	10	517
	Partial Frequency	1	1	1	11	1	1	1									

Normal cycle time: 10,040 TMUs

Allowance per cycle: 11.3%

Standard cycle time: 11,171 TMUs

Calculation of savings and break-even point

Difference between existing and proposed standard cycle times: 524 TMUs/work cycle

Difference between existing and proposed normal cycle times: 520 TMUs/work cycle

Hourly labor cost: $20

Operational savings for 8-hour workday: $7.40

Savings in allowances for 8-hour workday: $0.10

Total savings for 8-hour work day: $7.50

Invested cost: $25

BEQ: 0 workdays

[a] Improvement 5: Inclining work surface 20 degrees to impose safe working posture in backboard mailing operation.

allowances for backboard nailing operations. Our suggestion to incline the work surface requires a $25 investment that will never be paid off because it lengthens the standard cycle time by 524 TMUs.

Improvement 6 is similar to Improvement 2 and includes purchasing a sling to overcome the weight of the nail gun. As in Improvement 2, the cost is $150 and the result will be less stress in nailing operations, expressed as a 2% allowance reduction for nailing long boards to short boards, nailing support rods, and nailing backboards. A total of 81 TMUs per cycle will be saved, producing a savings of $1.23 per workday that will return the investment in about 122 workdays.

Finally, Improvement 7, which is operational in nature, recommends a separate pneumatic installation for the screwdriver and nail gun. In the existing situation, the worker switches the pneumatic power, alternating the nail gun with the screwdriver and vice versa twice per cycle. The new installation is estimated to cost about $350 and will reduce the cycle time by 220 TMUs (each tool change requires 110 TMUs). This improvement will result in a $3.58 reduction in cost per workday and will return in 98 workdays.

1.4   Discussion

The proposed improvements in a case study involving the assembly of a bed linen box are summarized in Table 1.4. Management can act to implement improvements through an initial investment and expect a return on their investment. The tables show that the required investment is not the only factor to consider when implementing an improvement. Improvements to operation times and decreased operator stress (expressed as reduced fatigue allowance times) may also influence decisions about implementing improvements.

The return on investment was calculated by analyzing investment versus time savings and using an hourly cost to calculate final results. The formulas we used in this chapter consider the cost of an ergonomic factor through allowances for the work cycle; work cycle times are usually reduced by ergonomic improvements. In cases where ergonomic improvements require capital investments and decrease standard throughputs, the company will never reach the break-even point as shown for Improvement 5. Although ergonomic improvements require fewer body motions and foster better posture, they may also call for longer action distances or idle time (waiting for machinery to perform). When these offsets occur, management must understand that making an ergonomic improvement is likely to pay off over the long run through direct financial savings and other benefits.

Good ergonomic operations that lessen physical stress and fatigue improve the well-being of employees, leading to long-term healthy, productive, and efficient work relationships. Good ergonomics practices also

Table 1.4 Summary of Comparison of Existing and Improved Metrics for Improvements 1–7 of Case Study

	Existing Work Situation	Improvement 1: Vertical Screwdriver	Improvement 2: Sling for Screwdriver	Improvement 3: Nailing Pronation	Improvement 4: Vertical Nail Gun	Improvement 5: 20-Degree Inclination of Work Surface	Improvement 6: Sling for Nail Gun	Improvement 7: Extra Pneumatic Installation
Normal cycle time (min)	5.71	5.72	5.71	5.71	5.71	6.02	5.71	5.58
Allowances (min)	0.68	0.65	0.66	0.67	0.64	0.68	0.63	0.67
Standard cycle time (min)	6.39	6.37	6.37	6.38	6.35	6.70	6.34	6.25
Savings per workday	—	$0.37	S0.39	$0.14	$0.84	–$7.51	$1.23	$3.58
Investment	—	$250	S150	$0	$ 250	$25	$150	$350
Workdays to return investment	—	669	0	0		Never	122	98
Manufacturing quantity to return investment	—	50,403	0	0		Never	9259	7504

reduce injuries and work-related disorders. Although these benefits are difficult to measure, they can appear as reduced costs for training new employees to replace injured veteran employees. Management will also save money that might have been spent defending lawsuits arising from injuries and paying insurance premiums. These factors can be calculated as the probability of injury (between 0 and 1) multiplied by the costs of injuries (lost work days, legal expenses, etc.). When the ergonomic factor is quantitatively expressed and derived through replicable measurement, it can be controlled by management and exploited for making decisions regarding workforce selection because it accounts for resting times and work area conditions.

Many companies do not include physical stress resulting from poor ergonomics in their standard time calculations. This is often the case in countries that are not developed industrially and labor costs are low. We would like to see the results (labor costs and break-even data) in such operations after good ergonomics practices become part of the equation. It is likely that in such cases some ergonomic improvements will not return their investments within a reasonable planning horizon because of lower savings per work cycle without considering improvements to allowances.

Just as labor costs vary among countries, ergonomic attitudes, policies, regulations, and union requirements may influence allowances set for standard cycle times. Changes in allowance percentages are reflected in standard cycle times. In order to distinguish between operational and ergonomic savings, an averaged allowance per cycle must be calculated and compared separately. If resting is not accounted for, the savings will cover only operational improvements. In this example, the improved versus proposed normal work cycle time is 7% improved where the standard cycle time is 8.1% improved, meaning that improving work conditions produced additional accountable improvements.

Without following the steps detailed in this chapter, it is difficult to make decisions based on viewing work activities to make the best analysis possible and designing improvements based on a quantitative analysis ranking tasks, improvement costs, and other factors. As shown in Elnekave and Gilad (2006), a standard time for an existing work situation can be set rapidly and remotely. Nevertheless, no formal techniques are available for formulating proposed work situations. We stress the need to validate how well the proposed work situations will reflect reality because the extent to which reality was reflected accurately was not studied.

Although PMTS techniques are known for their consistency and accuracy in prediction, it is often the analyst's task to predict correct work sequences, body motions, and working postures (Delleman 1999). Accuracy in predicting manufacturing times and costs is a requirement for strategic investment planning. Development of ergonomic models for analysis and redesign is time consuming and often of questionable value.

An experienced analyst can seek effective solutions intuitively and very quickly. The projected impacts of ergonomic improvements on an operation time must be calculated using PMTS to facilitate managerial decisions about improving existing methods.

Our approach does not compromise on this step and pushes toward quantitative analyses using simple computerized tools that will produce results quickly and remotely. Our approach includes combining work measurement calculations and ergonomic analyses to achieve a complete work analysis (Laurig et al. 1985, Gilad 1995, Laring et al. 2002). This approach proposes quantifying the measured posture and coupling operation times derived by formal work measurement techniques. Then the ergonomic factor can be expressed as a direct cause of the hidden trauma. An analyst can act upon that result by projecting motion sequences of various solutions, calculating their costs, and selecting the best one.

References

Ben-Gal, I. and Bukchin, J. 2002. The ergonomic design of workstations using virtual manufacturing and response surface methodology. *IIE Transactions*, 34, 375–391.

Delleman, N.J. 1999. Working postures: prediction and evaluation. PhD Thesis, TNO Human Factors, Soesterberg.

Elnekave, M. and Gilad, I. 2006. A rapid video-based analysis system for advanced work measurement. *International Journal of Production Research*, 44, 271–290.

Gilad, I. 1995. A methodology for functional ergonomics in repetitive work. *International Journal of Industrial Ergonomics*, 15, 91–101.

Gilad, I. and Elnekave M. 2006a. Time-based approach to obtain quantitative measures for ergonomics hazard analysis. *International Journal of Production Research*, 44, 5147–5168.

Gilad, I. and Elnekave M. 2006b. Inserting cost effectiveness into the ergonomic equation when considering practical solutions. *International Journal of Production Research*, 44, 5415–5441.

Laring, J., Forsman, M., Kadefors, R. et al. 2002. MTM-based ergonomic workload analysis. *International Journal of Industrial Ergonomics*, 30, 135–148.

Laurig, W., Kuhn, F.M., and Schoo, K.C. 1985. An approach to assessing motor workload in assembly tasks by the use of predetermined-motion-time systems. *Applied Ergonomics*, 16, 119–125.

Saleem, J.J., Kleiner, B.M., and Nussbaum, M.A. 2003. Empirical evaluation of training and a work analysis tool for participatory ergonomics. *International Journal of Industrial Ergonomics*, 31, 387–396.

Zandin, K.B. 1990. *MOST Work Measurement Systems*, 2nd ed. New York: Marcel Dekker, pp. 10–12.

chapter two

Workstation redesign, workplace improvement, and productivity

Reem Al Kadeem

Contents

2.1 Introduction

A workplace should be designed to enable employees to perform their jobs effectively. To achieve this, the workplace designer should keep two design factors in mind. The first is the large variability in sizes of people in the workforce population. The second is the need to understand the user population's culture, education, training skills, attitudes, physical and mental capabilities, and other characteristics.

The basic requirement for adequate workstation design is the anthropometric fit of a worker, the equipment or furniture, and the assigned task (Bridger 2003). In the absence of anthropometric data, engineers, designers, and facilities staffs can avoid anthropometric mismatches through a participative approach by which a panel of workers and the design team evaluate furniture prior to purchase. Fitting trials can be carried out to identify mismatches and develop requirements for accessories such as footrests, document holders, and other items.

The expectation was that both the quality of working life and productivity of operators would increase from the use of ergonomically designed furniture (Johnson 1985). A strong relationship exists between worker comfort and productivity. Unfortunately, this relationship has not yet been accepted by many industrial organizations whose managers assume that productivity and product quality are functions only of pay rates. This is an indication of a lack of understanding of the concepts of ergonomics and the benefits of incorporating ergonomics into the design of an effective workplace.

The goal of ergonomics is not just to reduce effort; it is rather to maximize worker productivity at a level of effort that is not harmful to the worker. For this reason, an ergonomics solution should not be the sole factor in deciding how much compromise should be made between comfort and efficiency. A solution should attempt to minimize the incompatibilities between the capabilities of a worker and the demands of his or her job (Tayyari and Smith 1997).

The following sections describe a study conducted by the Ergonomics team of the Faculty of Engineering, at two food factories. The work proceeded in three phases. The first was a survey conducted in two food factories to evaluate the levels of awareness and application of ergonomics guidelines and to quantify the exposure of workers to physical overexertion. The survey was also concerned with the incidence of work-related musculoskeletal disorders (WMSDs) in the two factories. The second phase was prioritization of problems presenting higher risks. The third phase was introducing ergonomic solutions by redesigning workstations or providing new equipment and then examining the impacts of the changes. Four workstations were reported as high risks for WMSDs of the upper extremities, lower extremities, and low back. The new workstations exhibited improved performance and reduced risk factors for WMSDs.

2.2 Unilever case study

2.2.1 Background

The food industry is considered relatively safe based on the low rates of traumatic industrial accidents. However, the hidden hazards that accumulate over time and cause WMSDs cannot be overlooked. A properly designed workstation is essential to workers' comfort, job satisfaction, motivation and sense of accomplishment—and hence their productivity.

WMSDs often develop from physical mismatches between workers and manual tasks they perform. Such mismatches introduce risk factors to the workplace and threaten worker well-being and performance. Factors such as repetition, force, unnatural posture, and vibration are associated with higher rates of injury and consideration of a workstation alone does not provide a key to understanding these injuries.

Growing evidence indicates that other factors are linked to injuries (Eaton et al. 2005). Working posture is an important factor in WMSDs (Bhatnager et al. 1985, Klaus et al. 1987). Those who stand while they work or work in constrained postures are more likely to be exposed to other physical demands (Tissot et al. 2005). In many research studies, ergonomic changes made in different aspects of workstations proved the influence of workstation design on worker posture and consequently on WMSD incidence, worker performance, and productivity (Sillanpää et al. 2003, Carrasco et al. 1995).

A fully adjustable ergonomically designed workstation was developed by Shikdar and Al Hadhrami (2007) with special features such as a motorized table to handle upward, downward, and angular movements; an ergonomic chair with adjustable seat pan, arm support and back support; and a mechanism for bin and tool locations.

The improved design could function as a sit, stand or sit–stand workstation. The range of table height movement was 700 to 1050 mm, with an angular movement of 45 degrees. Chair height was 430 to 560 mm, back-support tilt was 30 degrees, and elbow support height was 230 to 330 mm. The designers noted a significant difference in workstation set-up parameters expressed by participants and fixed parameters of existing assembly workstations.

The performance of the participants was 42.8% higher for the smart workstation in comparison of performance at the existing workstation. Flexibility in workstation set-up can eliminate the anthropometric and ergonomic problems of fixed workstations and boost operators performance.

Management and labor organizations are becoming increasingly aware that ignoring ergonomics principles in industrial workstations can lead to musculoskeletal harm and consequent reduced human performance. However, the application of ergonomic approaches in industry continues to be underestimated in developing countries. Many managers fail to consider the hidden costs in their analyses of the financial considerations of workplace injuries. For many managers, the costs of implementing ergonomic solutions far outweigh the potential benefits (McLean and Rickards 1998).

This chapter discusses attempts to implement ergonomic approaches in workstation redesign in two food packing factories of Unilever Egypt. The resulting solutions are simple modifications that may help reduce WMSD risk factors and lead to significant improvements in worker comfort, health, and productivity. The subsequent sections begin with risk assessments of the two factories under study, details of the suggested solutions, and examination of their effect on operators.

2.2.2   Risk assessment

A survey was conducted to assess the risk levels faced by workers in two Unilever food factories. The risks attributed to manual material handling are not covered in this chapter; they are examined and evaluated in another research work. This section focuses on jobs that involve highly repetitive movements and sustained loading that impact workers.

The survey involved three forms of data collection. The first step was compiling National Institute of Occupational Safety and Health (NIOSH) checklists based on workplace evaluations of musculoskeletal disorders. These checklists provide data about ergonomic risk factors and indicate whether a further analyses are required. The second form of data collection consisted of direct measurements and observations of the workstation features and dimensions. The final form was the use of the Corlett and Bishop scales of body parts and whole body discomfort (Corlett and Bishop 1976).

Videotaping was another survey tool used to monitor job procedures and the postures adopted by operators as they performed their jobs. Medical records and histories of work-related surgeries were also surveyed.

2.2.2.1   Worksite description

The survey covered 18 workstations in 10 workplaces. Questionnaires were administered to 54 operators (40 males ranging in age from 19 to 36 years and 14 females ranging in age from 18 to 32 years). All operators worked 8-hour shifts and their tenures with the company spanned from 6 months to 7 years. The workstations studied were involved in packaging, inspection, and sorting operations.

2.2.2.2   Workstation features

The observations and measurements concerned thorough evaluations of the workstations (seats and tables, lighting, and environmental conditions including noise and ambient temperature). No measurements were made of these conditions. Analysts concluded that:

- Workstation design was not based on any standards and failed to offer minimum baselines of comfort and support.
- Most jobs designed for performance in a standing position could be performed while seated if a proper seat design was introduced.
- Most machines and material handling equipment (conveyors) were not designed to suit Egyptian anthropometric dimensions; this made the selection of other workstation components very difficult.

Table 2.1 Operator Complaints and Possible Causes

Complaint Description	Percentage of Complaining Operators	Possible Cause(s)
Arm pain	44	Excessive work surface height relative to elbow level for seated operators
Neck and shoulder pain	82	Improper work surface height for seated and standing operators; lack of back support for seated operators due to large seat pan length that prevented operators from using backrests
Wrist pain	19	Repetitive hand movements of packers; sharp edges of work surfaces; insufficient room for wrist support
Lower back pain	94	Improper back rest position; large seat pan length that prevented operators from accessing job area; poor seat design that forced operators to stand throughout their work times
Foot numbness	92	High seat pan height required seated operators to dangle their legs
Leg and foot tingling	50	Prolonged standing during working hours

- In many workplaces, the cluttered layouts prevented the smooth flow of materials.

2.2.2.3 Worker complaints

The assessment questionnaire was developed to investigate possible complaints about pain in various body areas. Another form involved indirect evaluation of workstation design and the impact of the design on operators. By relating the results of measurements and observations of each workstation and operators' complaints, possible causes of the complaints were suggested as shown in Table 2.1.

2.2.2.4 Medical records

The factories' medical records revealed 24 cases of varicose veins among the standing operators and 6 cases of ganglion cysts out of each 12 were reported among 12 operators whose jobs involved repetitive wrist movements. Also, medical records revealed that several operators had back problems but no evidence related these problems to their jobs.

2.2.2.5   Rationale for change

The survey results revealed WMSD risk factors at all workstations and an obvious mismatch between workstation dimensions and the operators' anthropometric dimensions. This mismatch forced operators to adopt awkward postures and not use the available seats. Moreover, poor workstation designs and inconvenient layouts caused them to perform redundant tasks.

Several meetings were held with Unilever Egypt's Steering Committee and the most urgent cases were selected for further study. The suggested modifications were made with one condition in mind: not to replace machines or equipment. In other words, solutions were aimed to be simple and low cost. The next section discusses the selected workstations, the suggested modifications, and their effects on operator performance and discomfort.

2.2.3   Results of assessment

2.2.3.1   Sorting and packing workstations

One factory housed six sorting and packing workstations operated by six female operators. Their job was sorting tomato paste pouches, then packing every 24 pouches into a box, placing eight boxes in a single carton, and temporarily storing cartons at their workstations until they were transported to the next work area. Each operator had to pack 70 cartons daily. They first had to inspect pouches for date, seal integrity, and appearance, then the correct number of boxes in the carton in proper positions. Sorters worked two continuous 4-hour shifts separated by 30-min rest periods.

The workstation design included a worktable that was too high (80 cm) and the seat pan height was 51 cm. The height of the worktable caused the operators to lift their shoulders upward and forearms flexed excessively. Some of operators placed cartons on their seat pans to compensate for the difference between worktable and seat heights. The worktable design prevented operators from comfortable use of their spaces. They had to lean forward to reach the pouches, leaving their seat backrests unused.

WMSD incidence — The primary problems were lower back disorders and shoulder-girdle overuse syndrome. Medical records of operators reported six cases of ganglion cysts among 12 operators, three of which required surgeries within a year.

Workstation redesign — One proposed solution was to design a workstation based on Egyptian anthropometric dimensions (Al Kadeem 1996). The suggested design is shown in Figure 2.1. A prototype design was developed and examined. A pilot study involved the operators of only one sorting workstation (six females ranging in age from 17 to 26 years). Performance and discomfort measurements were used to evaluate the prototype design.

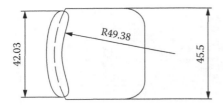

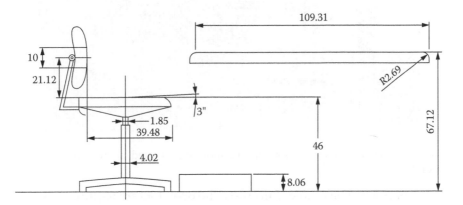

Figure 2.1 Dimensions of suggested workstation design.

The operators were asked to perform their sorting at a normal pace and observed continually for 15 min. They were videotaped to record their adopted postures. The Corlett and Bishop scale of body part discomfort (1976) was used to assess the perceived discomfort at the end of a 4-hour shift of continuous sorting. Working rates and error percentages (defective patches sorted as defect-free) were analyzed as performance measures for the current and redesigned workstations.

Improved performance and perceived discomfort — To examine the significance of the changes arising from the improved design, an analysis of variance (ANOVA) study was performed on the performance and discomfort measures recorded for the current and improved workstations. The result showed a significant decrease in perceived discomfort by 80% ($p < 0.01$). Working rate significantly increased by 40% ($p = 0.0$) when error percentage significantly decreased ($p = 0.0$).

2.2.3.2   Pouch conveyor workstation

Operators grasp pouches from a conveyor 90 cm high, place them in a carton, and temporarily store them on pallets. After filling a carton, an

Figure 2.2 Pouch conveyor workstation showing adopted posture by standing operators.

operator places it on a pallet 1 m away. The operators are required to stand continuously for 8 hr with only one 30-min break. Figure 2.2 shows operators at the station. The photo also shows the plastic boxes that operators used to raise the cartons to waist level to avoid bending their trunks while filling cartons.

WMSD incidence — Because the operators must stand for prolonged periods, blood and synovial fluids tend to accumulate in their legs and cause swelling and varicose veins. Also, the operators must keep their necks bent and experience pains of the neck, lower back, and legs.

Workstation redesign — The suggestion was made to allow operators to sit–stand while performing their jobs instead of standing. A sit–stand seat based on conveyor height was designed. The seat allows operators to alternate between sitting and standing positions. Also, the conveyor was modified to support each carton as it is filled. Figure 2.3 shows the dimensions of the sit–stand seat designed to accommodate the conveyor height and enabling the average operator to keep his forearm flexed 90 to 110 degrees to his upper arm. The designers utilized Egyptian male anthropometric dimensions since all operators at this station were males. The modifications were examined, and discomfort levels and work rates measured for both existing and proposed designs.

Improved performance and perceived discomfort — ANOVA was performed on both working rates and worker discomfort data. The

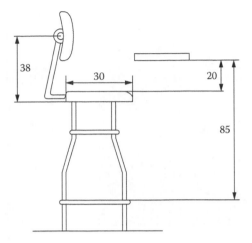

Figure 2.3 Sit–stand seat dimensions for 88-cm conveyor.

analysis showed a significant decrease in perceived discomfort of the legs and lower back region ($p < 0.05$) and no significant variations in working rate, possibly because this station is machine paced.

2.2.3.3   Tea bag packing workplace

The 40 tea bag machines in the factory were staffed with 2 operators at each station. Each operator temporarily filled cartons boxes of tea bags. They transported each carton weighing 15 kg to a gluing and coding station 40 m from the most distant filling machine and 10 m from the nearest one. Cartons had to be refilled after coding. Figure 2.4 depicts the old and

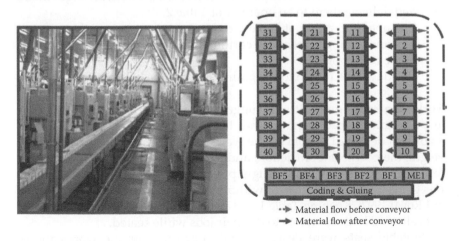

Figure 2.4 Left: Installed conveyor. Right: Before and after material flow.

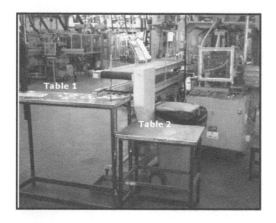 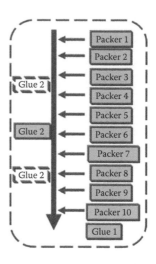

Figure 2.5 Left: Old workstation. Right: Packing machine layout.

new material flows. At left is a conveyor installed between two lines of tea bag machines.

The packing machines shown at left in Figure 2.5 are 1.7 m from the conveyor responsible for transporting packed cartons. At right, the figure shows tea packet packing (RT) workstations; the thick arrow represents the conveyor. Two operators at each station perform the following steps:

- After cartons are folded and glued on the bottoms, they are fed to packing machines.
- The standing operator at Table 1 receives tea packets, arranges them, and delivers them to an operator at Table 2.
- The operator sitting at Table 2 receives packets, wraps them in cellophane, arranges them in a carton, then transports finished cartons to the conveyor.
- Filled cartons return to the gluing machine to be sealed.

Workplace redesign — It was suggested to install a supply chute conveyor to link the packing machines to the end conveyor. The chute conveyor was designed to convey the weight of the carton. Based on carton weight, the coefficient of friction, and the angle of inclination, a detailed design was developed. Only a slight push is required to accelerate the cartons to meet the main conveyor without plugging at impact points (Stuart and Royal 1992). Keeping Table 1 and providing the workstation with two seats allowed both operators to do their jobs while seated.

Suitable seats were designed and suggested for the workstations. Another modification was to add another gluing machine (Glue 2) that

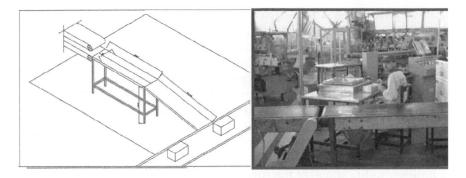

Figure 2.6 Left: Supply conveyor. Right: Operators' suggested enhancements.

moves to each station to glue cartons. Figure 2.6 illustrates the suggested conveyor design. The only new equipment required for the modification was a movable gluing machine. This modification enhanced the packing rate. The conveyor could not be implemented or examined for this study because this workstation is undergoing further automation. Management was reluctant to complete any other improvements at this station. However, the operators modified the design by moving their worktables next to the main conveyor to avoid the need for frequent sitting and standing.

A conveyor was installed to serve the whole group of tea bag packing machines including the coding and gluing machines. The improvement yielded reductions in packing time and worker exertion. In the course of the motion and time study, operators' physiological parameters (blood pressure, pulse rate, and Borg measurement) were checked after 4 hours of work and the results used to assess the improved working conditions.

Improved performance and physical effort — Measurements were recorded for six male operators as they worked under existing conditions and after the improvements were made. ANOVA was used to assess the statistical significance of variations in performance and physical measures. The results showed a significant change in pulse rate ($p = 0.04$) and a significant (65%) time reduction ($p = 0.00$). A significant (53%) increase in packing rate was also proved ($p < 0.01$). A Borg scale showed a 30% reduction in physical effort with $p = 0.0$.

2.3   Discussion

While the ergonomics discipline has been practiced in industry for years, its application in developing countries is not as common. The role of comfort in enhancing efficiency is always undervalued in some parts of the world.

This study discusses an attempt to introduce ergonomics into Egyptian industry. Although the food packing industry is considered rather safe, a risk assessment revealed risk factors for WMSDs.

The redesigned workstations based on the Egyptian anthropometric dimensions significantly reduced perceived discomfort. Moreover, properly designed seats enabled operators to perform their jobs in sitting and standing positions instead of standing constantly during their work shifts.

In the tea bag production workplaces where layouts were redesigned, the suggested solutions reduced unnecessary work. Workers had to temporarily fill cartons for transport to the coding and final packing stations. Installation of a conveyor eliminated this task, resulted in productivity improvements, and decreased physical exertion. In one case, only part of the suggested plan could be implemented. Management was reluctant to complete modifications of the supply conveyor and seats. However, production rates still improved. Workers were enthusiastic about the introduction of ergonomics to their workplace and suggested moving their stations next to the conveyor. This eliminated the frequent standing and sitting movements required earlier.

Operators are directly affected by ergonomics improvements. In addition to understanding ergonomics guidelines (Carrasco et al. 1995), they must be involved in determining the tasks that pose physical risks and must also test the modifications.

References

1. Al Kadeem, R.M. 1996. A Study of the Effect of Computer Workstations on Operators' Errors. M. Sc. Thesis, Alexandria University, Egypt.
2. Bhatnager, V., Drury, C.G., and Schiro, S.G. 1985. Posture, postural discomfort, and performance. *Human Factors*, 27, 189–199.
3. Bridger, R.S. 2003. *Introduction to Ergonomics*, 2nd ed. London: Taylor & Francis.
4. Carrasco, C., Coleman, N., and Healey, S. 1995. Packing products for customers: an ergonomic evaluation of three supermarket checkouts. *Applied Ergonomics*, 26, 101–108.
5. Corlett, E.N. and Bishop, R.P. 1976. A technique for assessing postural discomfort. *Applied Ergonomics*, 19, 175–182.
6. Eaton, J., Kerr, M., and Ferrier, S. 2005. Dealing with work-related musculoskeletal disorders in the Ontario clothing industry. http://www.wsib.on.ca/wsib/wsibsite.nsf/public/researchresultsmusculoskeletal
7. Johnson, S.L. 1985. Workplace design applications to assembly operations. In Alexander, D.C. and Pulat, B.M., Eds., *Industrial Ergonomics: A Practitioner's Guide*. Norcross, GA: Institute of Industrial Engineers.
8. Klaus, M.B., Dickens, K., and Pappas, P. 1987. Postural discomfort evaluation of production workers. In Asfour, S.S., Ed., *Trends in Ergonomics and Human Factors IV*. Amsterdam: Elsevier, pp. 655–662.
9. McLean, L. and Rickards, J. 1998. Ergonomics codes of practice: the challenge of implementation in Canadian workplaces. *International Journal of Forest Engineering*, 1, 55–64.
10. NIOSH Elements of Ergonomics Programs-Toolbox Tray 5-A. http://www.cdc.gov/niosh/eptbtr5a.html

11. Shikdar, A. and Al Hadhrami, M. 2007. Smart workstation design: an ergonomic and methods engineering approach. *International Journal of Industrial and Systems Engineering*, 2, 363–374.
12. Sillanpää, J., Nyberg, M., and Laippala, P. 2003. A new table for work with a microscope: a solution to ergonomic problems. *Applied Ergonomics*, 34, 621–628.
13. Stuart, D. and Royal, T.A. 1992. Design principles for chutes to handle bulk solids. *Bulk Solid Handling*, 12, 447–450.
14. Tayyari, F. and Smith, J.L. 1997. *Occupational Ergonomics: Principles and Applications*. London: Chapman & Hall.
15. Tissot, F., Messing, K., and Stock, S. 2005. Standing, sitting and associated working conditions in the Quebec population in 1998. *Ergonomics*, 48, 249–269.

chapter three

Productivity and worker age

**Kurt Landau, Margit Weißert-Horn,
Angelika Presl, and Regina Brauchler**

Contents

3.1   Aging workforces

Many industrial countries now face the need to adjust to the phenomenon of aging workforces and are aware that this aging trend will accelerate in the years to come. Forecasts show that the 50 to 64 age group will have ousted the 35 to 49 age group from its current leading position of number of employed in industrial countries by 2020.

We should note, however, that no clear definition of older workers exists; we have only somewhat objective criteria covering the group. The literature contains widely differing opinions on the age at which a person can be classified as an older worker.

According to the definition of the United Nations' Organisation for Economic Co-operation and Development (OECD), older workers have reached the second half of their professional or occupational careers. The

U.S. Bureau of Labor Statistics (1997) classifies all employed persons over the age of 55 as older workers and states that 40% of people in this age group are employed. U.S. workers in the 65+ age group tend to switch from part-time to full-time work. About 65% of those who continue to work beyond the age of 65 have full-time jobs. The Bureau of Labor Statistics is also expecting an increase in employment of over 80% in the 65 to 74 age group through 2016 as a result of inadequate levels of pension income.

The overall employment rate of older workers (55 to 65) in the European Union was around 44.7% in 2007 (Puch 2009), but the picture is far from uniform in the individual countries. While southern and southeastern Europe are experiencing dramatic levels of population shrinkage (with all the accompanying negative demographic consequences), population decline in central Europe is proceeding more slowly and steady state prevails in Scandinavia. In contrast, Turkey and Ireland report population growth. Factors influencing trends in workforce age include birth rate, life expectancy, and net migration rate.

Figure 3.1 shows employment rates by age group in selected industrial countries. Japan and Switzerland have the highest rates in the 55 to 64 age group and Austria has the lowest. The Austrian figure is influenced by that country's heavily subsidized early retirement program.

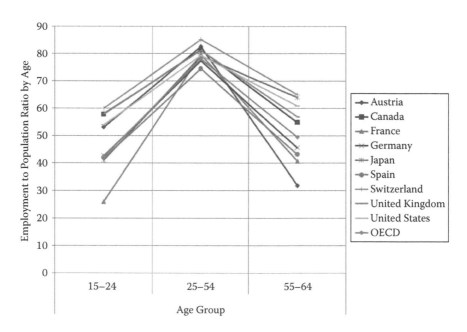

Figure 3.1 Employment-to-population ratio by age in 2005. (Source: OEDC, 2006. *Employment Outlook*. www.oecd.org/els/employmentoutlook/EmO2006.)

Strong media focus on demographic change plus increasing concern about its consequences on the part of European politicians enhanced public awareness of its effects:

- Draining of the domestic labor market, sinking economic growth, stifling incentive to innovate, intergenerational social conflicts, problems in financing of social security
- Dismantling of social security benefits despite increases in contributions.

Commenting on the problem of older workers in its guidelines for political action to safeguard employment rates, the Council of the European Union called on politicians in member countries to set employment targets within the EU and attain them as quickly as possible:

- Overall employment rate of 70%
- Female employment rate of 60%
- Older worker (55 to 64) employment rate of 50%

The council also urged cuts in unemployment and non-employment rates. Member states are expected to set their own national employment rate targets. The actions should:

- Channel more people into employment and keep them there, augment labor supplies, and modernize social security systems.
- Help make the industrial workforce more adaptable.
- Increase investments in human resources by improving education and vocational training.

Scientific and political debates on this subject frequently fail to differentiate employability and employment opportunity properly. *Employability* is definable as a person's ability to perform a given task at a given time based on his or her health status, professional or occupational skills, degree of motivation, and attitude. *Employment opportunity* is the essential condition for finding a job for a person who is able and willing to work. It depends on the labor market's capacity to absorb available labor and the ability and willingness of industry to retain its existing workforce and recruit new workers. Thus, enhancement of employability is not per se an adequate method of achieving higher employment rates for older workers. One essential element is introducing official regulations and agreements on wages and working conditions to increase employment opportunities in industry.

3.2   Age and capabilities

We differentiate calendar, biological, functional, and social ages. Social age is the age for which a person's status and esteem in society qualifies him. Biological age has shown a sharp drop relative to calendar age in recent years. Today's 70-year-olds have the same biological age that 60-year-olds had in 1970. People who reach age 65 today statistically still have a quarter of their lives before them. In contrast and paradoxically, social age has risen relative to functional age.*

For this reason, calendar age is not, as still frequently assumed, a reliable indicator of physical and mental capabilities. The effect of aging on professional and occupational capabilities must be viewed against a person's overall history, i.e., aging must be seen more as a continuous individual process than as a phenomenon suddenly manifesting itself at a specific number of years.

The aging process certainly involves physical changes, but also means changes in mental and emotional attitudes and social behaviors. Although powers of endurance start to wane from age 40 onward, regular training can block or slow this decline for a considerable period. It is reasonable to assume that regular physical work would be equally effective in slowing the aging process. Muscular strength suffers a roughly 30% decline between ages 30 and 65, but is largely attributable to atrophy of skeletal muscle through inactivity. This loss can also be countered to a considerable extent by appropriate training. It is clear that this type of training can benefit physical condition and coordinative capability. Even more remarkable than the effects of physical training on physical capabilities is the evidence that endurance training can prevent or at least abate the creeping decay of cognitive functions normally accompanying advancing age.

The wide range of sensory, motor, and cognitive deficits that appear in elderly and old people are well known and clearly demonstrated. It this seems reasonable to expect a similar decline in work performance with advancing age, but this is not necessarily the case. Recent investigations, particularly longitudinal Finnish studies (Ilmarinen et al. 1997) proved that work capability is statistically not age-dependent. It tends to stabilize and even in some cases increase with advancing age. Older workers:

- Know the intuitive details of work processes.
- Understand a company's work structures, market demands, and customers.
- Are willing to work hard despite their health problems.

* We do not intend to make an extensive review of current research here. Reference can be made to the synopses of Sanders (2009) and Skirrbek (2003).

- Know how to handle customers.
- Are often very flexible, because they are no longer hampered by family duties.
- Are highly motivated and loyal.
- Exhibit high levels of staying power.

Furthermore, their lengths of service with a company are calculable (Adenauer 2002). This list of positive points effectively contradicts the common assumption that performance automatically declines with advancing age. It must be noted, however, that an employee's so-called reserve capacity (margin between actual work input and stress levels on one hand and his or her maximum performance capability on the other) tends to decline with age.

The literature contains many analyses of the performances of elderly politicians, industrialists, and literary or artistic geniuses (Cf. synopsis by Marchetti 2002). Many of Johann Wolfgang von Goethe's most famous works were written in old age, in some cases when he was over 80. Pablo Picasso was still very active at age 80. Sir Winston Churchill was 76 when he became England's prime minister. Konrad Adenauer was 79 when he was elected Federal German Chancellor in 1949. He was re-elected three times and held the office until age 87. Mother Teresa dedicated herself to helping the world's poor until her death at age 87 and was awarded the Nobel Peace Prize when she was 69.

In an analysis of publications by Nobel prizewinners, the *Scandinavian Journal of Economics* found that, despite wide variations among several fields of science, many prizewinners were over 50 when they did their most important work. Film directors tend to stay productive in old age. For example, Michelangelo Antonioni did some of his best work at age 78.

3.3   *Job-linked physical performance and age*

Many published reports of mental and physical performance levels in old age are based on laboratory studies or the experiences of physicians and psychologists. There is a serious lack of studies of the capabilities and skills of industrial workers. It seems that the paradox of decline in physical performance despite high levels of productivity from all employees cannot be explained by investigation of body organ or physiological functions.

Three German rehabilitation clinics (Klinik Bavaria) conducted a study on a random sample of 4184 rehabilitation patients treated at their facilities between 2002 and 2006 to determine whether and to what extent job- and qualification-related performances of the patients were age- or sector-dependent. The Bavaria Rehabilitation Patient Assessment (BRA; Landau et al. 2007) was used to assess patient capability at the time of the initial diagnosis immediately prior to the start of rehabilitation. The

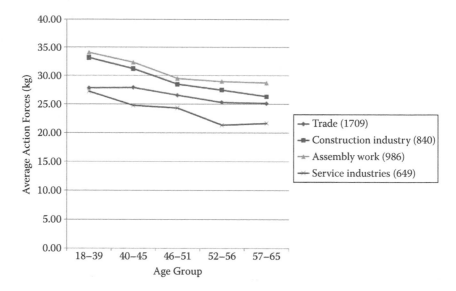

Figure 3.2 Lifting performance of rehabilitation patients from different sectors of industry as measured by EFL full and screening tests (n = 4184).

ratings were based on functional capacity evaluation (EFL) measurements (Isernhagen 1992) and devised by specialists in industrial or rehabilitation medicine. The patients were divided into four job-related categories:

- Trade: drivers, retail sector, accounts, textiles (n = 1709)
- Construction, agriculture, transportation (n = 840)
- Assembly work (n = 986)
- Services: catering, nursing, cleaning (n = 649)

The patient ages spanned 18 to 65 years and were spread more or less uniformly across the four categories.

Variance analyses based on the work sector and age factors revealed the capability levels of patients from each sector for performing lifting activities (Figure 3.2). Similar analyses of carrying, pushing, and pulling activities including tests involving forced postures were performed. The mean results reveal only a slight age-related decline in performance, and inter-sector variances are also very minor. Both foregoing factors make a contribution of only around 4% to explaining the overall variance. The results allow no conclusions about the extent to which *healthy worker effects* or job-related training effects are responsible for the results in the older age groups.

It should be noted that all the test subjects had been diagnosed with orthopedic diseases and that any extrapolation of the results to the workforce as a whole must therefore be interpreted with caution. Furthermore, it is not possible to make any deductions from individual case data because the analysts conducted a cross-section study. The value of these data lies in the fact that they qualify the opinion that low-performance workers will leave a company prematurely and that the remaining healthy worker effects will bolster productivity. Even workers diagnosed with orthopedic diseases suffer very few deficits in job-related capabilities.

3.4   Age, subjective symptoms, and disease

3.4.1   German trade union surveys

For some years now, German trade unions have been conducting detailed surveys of stresses to which workers are exposed and symptoms from which they suffer. The results revealed, for example, that about one-third of respondents working in jobs classified as "good" in the DGB Index doubt whether they will be able to continue performing their present jobs until pension age.

If one ignores the recent strong criticism of the validity of the DGB* Index (*German Journal of Ergonomics* 2010), the subjective opinion expressed by these respondents constitutes a de facto problem that must be factored into strain analysis and job design. According to the 2010 DGB Index, the German workforce works under widely differing conditions: 12% of workers have good jobs; 33% have bad jobs, and 55% have average jobs (Figure 3.3).

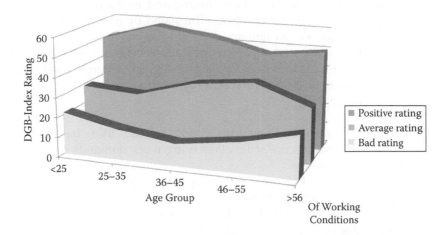

Figure 3.3 DGB Index rating spread across worker age groups. The standardized questionnaire survey covered 4150 respondents during the first quarter of 2010; the response level was 56.5%. (*Source:* International Institute for Empiric Social Economy, 2010.)

* DGB is a confederation of German trade unions.

Workers describing their jobs as stressful and monotonous frequently complain of musculoskeletal symptoms (Fuchs 2010). The more stressful their ratings of working conditions, the more frequent are the complaints of disease symptoms. For example, 55% of workers aged 50+ claimed to suffer more than eight different disease symptoms.

However, generalizations must be treated with caution. Statements relating to disease symptoms and working until pension age are multifactorial and strongly influenced by accumulations of physical and mental job stresses and demotivation resulting from lack of career development opportunities. The conviction that it will be possible to continue working until pension age increases in direct proportion to level of employee qualification and is most prevalent among persons fulfilling management functions. Company size is another influencing factor; the smaller the company, the more workers doubt their ability to stay until pension age (International Institute for Empiric Social Economy, 2008).

3.4.2   Automotive assembly line work

This section discusses a study involving a population performing more or less homogeneous work on an automobile assembly line. Two diametrically opposed concepts are involved: (1) older workers are phased out under early retirement schemes (before musculoskeletal problems arise) and preference is given to younger workers, in many cases on a temporary, subcontracted basis; and (2) ergonomic and medical risk points in assembly processes are identified and eliminated by modification of the work model regardless of worker age.

The results of a case study of 256 work stations on an assembly line for medium-priced cars manufactured by a globally active corporation were reported (Landau et al. 2008). The following methods were used to determine the workloads:

- Workload Screening Procedure (WSP; Schaub 2004) based on a quantitative point-score scale classifying work demands into three workload categories
- Ovako Working Posture Analyzing System (OWAS; Karhu et al. 1977) for assessing lifting and carrying activities
- BkA method (Schaub et al. 1999) for evaluating workloads caused by unfavorable body postures and lifting
- Ratings of three company experts (company medical officer, production planner from relevant production unit, and one ergonomist working at various production units)
- European Standard EN 614-1 covering ergonomic design of workplaces based on National Institute for Occupational Safety and Health (NIOSH) weight limits

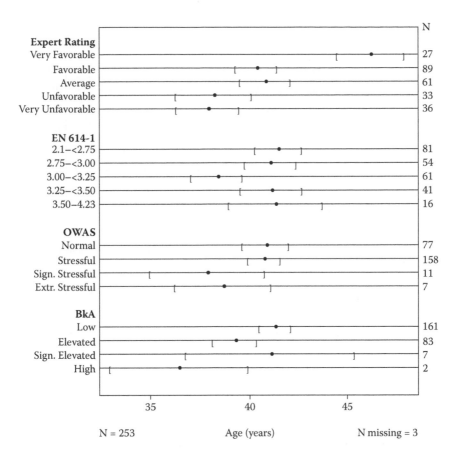

Figure 3.4 Mean age (± systematic error) of occurrence of various levels of work-loads based on determined by expert rating, OWAS, BkA, and EN 614-1. $N = 256$.

- NIOSH (NIOSH; Waters et al. 1994) for recommended weight limits (RWLs) in jobs involving application of high forces
- REFA* (REFA 1987) for evaluation of load handling

A significant correlation exists between ages and workload measurements obtained by the expert rating (Figure 3.4).[†] While OWAS and BkA evaluations reveal age differentiations similar to the expert rating, the EN 614-1 ratings reveal no distinct age dependency. Figure 3.5 shows the age distribution for presence or absence of musculoskeletal symptoms. The data reveal no correlations with worker age.

* German organization in work design, industrial organization and company development.
[†] Kruskal-Wallis test: $T = 12.38$, df $= 4$, $p \leq 0.015$. Linear association: $T = 7.36$, df $= 1$, $p \leq 0.007$.

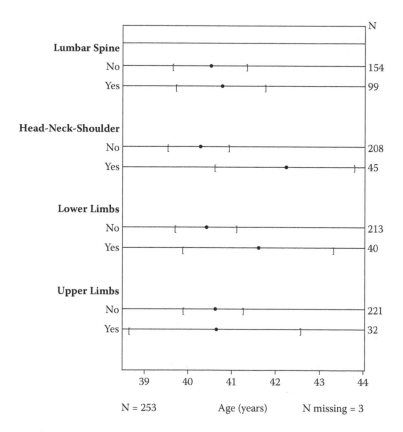

Figure 3.5 Mean age (± systematic error) of occurrence of musculoskeletal symptoms. N = 256.

We used binary partitioning analysis to discover which covariables are associated with the severity of two workplace target variables: (1) musculoskeletal symptoms and (2) subjectively perceived strain. The more unfavorable the expert rating of the job, the younger are the workers employed. This tendency is particularly pronounced for assembly line jobs rated very favorable that tend to be occupied by older (average age of 45) workers. It would seem that certain types of jobs are allocated to certain types of workers based on workload.

The very favorable jobs were given to older and smaller* workers and those tending to overweight (elevated body mass index [BMI]). Jobs were subjectively perceived as especially monotonous when they had high EN-614-1 ratings (≥3.25) and workers were older (>53 years). Younger

* Kruskal-Wallis test: T = 11.01, df = 4, p ≤ 0.026. Linear association: T = 4.82, df = 1, p ≤ 0.028.

workers tended to rate their assembly jobs as less monotonous and less overtaxing. Time pressure was rated somewhat higher by workers performing average, unfavorable, and very unfavorable jobs; lower ratings were noted by those with favorable and very favorable jobs. Satisfaction with the current job co-varies with the level of perceived strain. No significant effects of age on time pressure perception or dissatisfaction were detectable.

Age and length of service at the same job are covariables that may be expected to correlate with the presence of musculoskeletal symptoms. The higher the worker age and the longer the length of service in a physically stressful job, the more symptoms are to be expected. However, we already noted that any such association may be distorted by the practice of allocating certain types of job to certain workers.

Partitioning analysis revealed that older workers on average held favorable jobs and older workers with less favorable jobs suffered disease symptoms. Mentions of musculoskeletal symptoms in the head–neck–shoulder region were more frequent in workers of relatively short heights (≤ 164 cm) whose jobs rated ≥ 3.00 by the EN-614-method. Taller workers complaining of head–neck–shoulder symptoms were over 41 years of age and had jobs with EN-614-1 ratings ≥ 3.00 and expert ratings worse than favorable. Lower limb symptoms tended to occur more frequently in jobs with expert ratings of unfavorable or very unfavorable and in workers with heights of 181 cm or less.

Upper limb symptoms are mentioned more frequently in jobs rated by experts as unfavorable or very unfavorable and also at average or (very) favorable work situations involving underweight workers (BMI < 21). No significant effects attributable to age were detectable.

There was also almost certainly a healthy worker effect, i.e., many workers suffering from musculoskeletal symptoms had already opted for early retirement schemes, leaving a higher proportion of younger, healthy workers in their jobs (according to information received from the company's medical officer and the human resources department).

3.4.3   Musculoskeletal disease in the population of rehabilitation patients

The Klinik Bavaria study cited in Section 3.3 involved testing of performance in a population of rehabilitation patients. This section addresses only the musculoskeletal disorders occurring frequently in that population (classified in accordance with ICD-10-GM 2010):

Classification	Disorder
M17	Arthrosis of knee
M50	Cervical disk disorders
M51	Other intervertebral disk disorders
M53	Other dorsopathies not classified elsewhere
M54	Dorsalgia
M75	Shoulder lesions

These were the most frequent diagnoses in all age groups of the orthopedic rehabilitation patient population.

Job demands were recorded on an ordinal scale by specially trained medical staff using the Bavaria Rehabilitation Patient Assessment Method (BRA; Landau et al. 2007). The data were analyzed with six-digit tables and the results tested for asymptotic significance with a Pearson chi-square test. It should be noted that selection and information biases were present in the study population. The data are not suitable for extrapolation to the workforce as a whole because all the test subjects had been diagnosed for prior disease conditions.

Figure 3.6 shows the age profiles of the selected musculoskeletal disorders. The spike appears in the 45 to 54 age group. Several factors may explain the spike. First, the damage accumulated over a long period of service and now needs treatment. Second, the symptoms may serve as covers for burnout or resignation. In this case and especially with a diagnosis of unspecific dorsopathies, a patient may have exerted pressure on his doctor to prescribe a course of rehabilitation.

A number of reasons may explain the downturn in frequency of these diagnoses in the oldest age group. The first explanation is possible selection effects resulting from early retirements or transfers to other jobs of employees suffering from these disorders; this would raise the percentage of healthy persons in that age group. A second factor is reluctance of the social insurance institutions to approve rehabilitation courses shortly before retirement. Third, these older patients may have been transferred to less stressful jobs. Further analysis of this sample of rehabilitation patients shows:

1. The lower the employee's occupational qualification, the higher the risks of dorsalgia and dorsopathies.
2. A high degree of association exists between load-handling jobs and dorsopathy.
3. A clear association exists between forced postures and dorsalgia or dorsopathy.

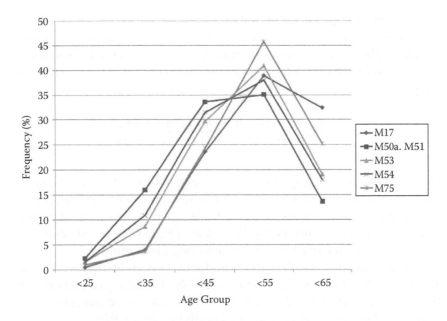

Figure 3.6 Age-specific frequency distribution of selected musculoskeletal diagnoses. M17 = arthrosis of knee. M50 = cervical disk disorders. M51 = other inter-
vertebral disk disorders. M53 = other dorsopathies not classified elsewhere. M54
= dorsalgia. M75 = shoulder lesions. $N = 3451$.

4. High hand forces tend to vary in causing both dorsopathy and
   shoulder lesions (Figure 3.6).

The association between load handling jobs and dorsopathy tallies with
numerous similar reports in the relevant literature (Cf. Walker 2000).

3.5 Productivity models

Even job-related performance tests yield no data on actual labor productivity at the workplace. The *productivity* term describes the yield obtained
from the sum of production factors: (1) labor, land, and capital (economist's
definition) and (2) manufacturing installations, raw materials, and labor
(business management definition). Consequently, the simplest method of
calculating labor productivity is to use the overall wage costs per unit.
Physical productivity (π_p) can be calculated as

$$\pi_p = \frac{Output\ in\ physical\ units}{Factors\ used} \tag{3.1}$$

Calculation of π_p makes sense only in the case of more or less standard, unchanging products or services. Comparing industrial sectors, manufacturing plants, work teams, or individual workers tend to yield inconclusive results for calculating π_p. Comparisons of monetary productivity are preferred.

$$\pi_m = \frac{\textit{Value of production output} \text{ [in monetary units]}}{\textit{Value of factors used} \text{ [in monetary units]}} \qquad (3.2)$$

However, the resulting quotients fail to make allowance for changes in wage costs following revisions of union wage agreements, technical progress, and increases in work intensity. Consequently, this form of productivity is aligned to monetary market values. Fricke (1961) used "monetary veil" to describe the masking factors affecting productivity.

Macroeconomic calculations of productivity of industrial work teams are often calculated with adaptations of Cobb-Douglas production functions (Goebel and Zwick 2009). Details of U.S. calculation methods can be found in a Bureau of Labor Statistics handbook (BLS 1997):

$$\ln\left(p_{j,t)}\right) \approx c + \beta \ln\left(k_{j,t}\right) + \sum_{i-\{0\}}(1-\beta)\left(\frac{\alpha_i}{\alpha_0}-1\right)\left(\frac{L_i}{L}\right)_{j,t} + \varepsilon_{j,t} \qquad (3.3)$$

where:
  p = value added per head.
  k = capital per head.
  i = age group.
  L_i = total number of employees in company L.
  α_i = marginal product of age group i.
  α_0 = marginal product of reference age group.

3.6   *Productivity measurements*

Boersch-Supan and Weiss (2005) classify the age-related economic measures of productivity into four categories of studies investigating:

1. Associations between a company's productivity at factory level and age of factory workforce
2. Individual wage rates as measures of individual productivity
3. Subjective appraisals of worker performance by their superiors
4. Direct assessments of individual worker productivity

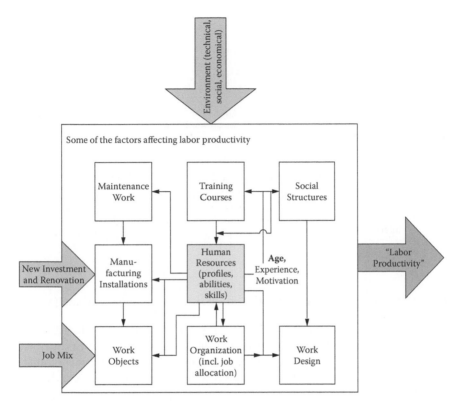

Figure 3.7 Factors affecting labor productivity.

Age-related measurement of *productivity at the factory level* is a highly aggregated procedure involving worker age and a whole raft of other factors, most of which cannot be controlled. Figure 3.7 lists these factors and highlights the fact that worker age, experience, and motivation are only three among a large number of variables influencing a manufacturing unit's labor productivity. Moreover, worker age and worker experience in a specific job are highly interdependent. Increases in labor productivity are certainly greater early in a worker's career than those seen in later years.

Younger workers will normally attend basic and advanced training courses that will enhance their productivity. Older workers are less likely to attend internal training courses. On the other hand, younger workers are not productive on workdays when they attend training courses. Furthermore, training effects are not particularly relevant when they involve simple production operations, for example, car assembly.

In addition to the age structure of a factory workforce, labor productivity can be indirectly affected by changes in capital productivity resulting from new investment and renovation of manufacturing installations.

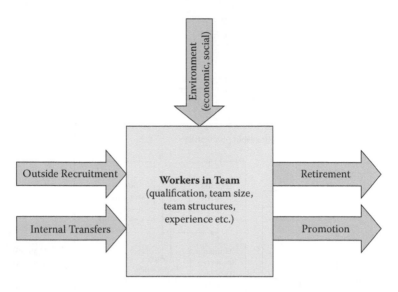

Figure 3.8 Changes in team structure that may distort productivity analyses.

Other significant factors are quality of work, factory organization, job design, maintenance of manufacturing installations, and last but certainly not least, the factory's social structures. The results of any investigation focusing solely on associations between worker age and labor productivity will therefore tend to be distorted by a large number of other factors and confounding variables.

Another important factor is the probability of a significant number of selection and healthy worker effects. Productivity data going back one or more years will likely be affected by continual changes in team composition resulting from recruitment of new workers or transfers to other work (Figure 3.8).

Factory size and industrial sector play significant roles. For example, staff composition in the information technology sector is totally different from that found in traditional engineering. Departures from the team are particularly important. It is possible that pensioning off older workers with severance payments will rid a team of some less productive members, but it is also possible that job-linked disease is a contributory factor in early retirement. At the other end of the scale, promotion is an important factor because it causes high-performance members to leave the team for management duties. Another possible confounding factor is a job description that fails to explain job demands clearly.

Although wage costs represent the simplest method of assessing factor input in financial terms [Equation (3.2)], wage levels across age groups yield little hard data on productivity. Wage payments are linked to type of job, worker qualification, individual performance, and seniority

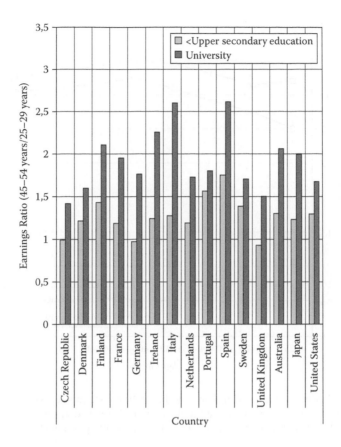

Figure 3.9 Earnings ratio by age group and level of educational attainment. (Source: OECD, 1998. *Society at a Glance: OECD Social Indicators.*).

supplements negotiated in union agreements. Research results on the interplay of these factors are very meager (Gelderblom 2006).

Figure 3.9 shows OECD statistics on remuneration level ratios between younger and older workers in various countries. The earning span between younger and older workers with lower educational qualifications ranges between 0.9 and 1.75 and is highest in Spain and lowest in the United Kingdom. The picture is radically different in employees with university qualifications, ranging from roughly 1.4 in the Czech Republic to over 2.6 in Spain and Italy.

It is obvious that this situation cannot be explained away by seniority payments alone. Age brings more responsibilities and more complex job demands that must be reflected adequately in remuneration levels.

Figure 3.10 shows data from a Dutch study (Ours and Stoeldraijer 2010) of a population of 44,371 workers in 13,941 companies that indicates (a low level

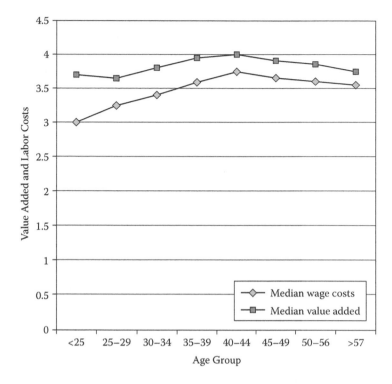

Figure 3.10 Median value added and median labor costs in successive worker age groups. (*Source:* Ours, J. C. van and Stoeldraijer, L. 2010. Age, wage, and productivity. Bonn: Institute for the Study of Labor, Discussion Paper 4765.)

of) correlation between median value added and median labor costs in successive worker age groups. The curves show only a vague hint of the inverted U frequently alleged to exist in many age-related studies. This appears to put Skirrbek's (2003) opinion that "younger workers are underpaid and older workers overpaid relative to their productivity" into perspective.

There is very substantial room for error in subjective appraisals of work productivity by superiors. These include the "images" attached to certain jobs, allocation of certain jobs to younger or older workers (allegedly) based on congruence of job demands and worker capabilities, preconceived opinions about decline of productivity, and many other factors. The number of causes of errors in productivity appraisals by superiors may be reduced by careful design, conduct of the appraisal, and also by re-tests (Avolio et al. 1990).

Direct measurements of productivity may be useful in longitudinal studies of individuals. Reference has already been made above to the productivity of older scientists and artistic geniuses, but similar data for industry and commerce workers are scarce. The manufacturing industry simply has no equivalent for a list of a composer's works or a scientist's publications. The few data on factors like work quality are filed in human resource department files and largely inaccessible because of data protection regulations.

3.7 Cross-section studies

Most econometric cross-section studies are based on aggregate data that treat a company or group of workers (not individual workers) as independent variables. Productivity is defined as an age group's indirect contribution to corporate productivity. Marginal productivity figures for different professions. occupations, age groups, genders, and other parameters are calculated, comparisons of remuneration are made, and opinions are expressed on relationships between productivity and remuneration (Aubert and Crépon 2003; Goebel and Zwick 2009).

Many cross-section studies demonstrate that, in contrast to work experience, worker age is not a reliable basis for forecasting performance of age groups. For example, a study by Avolio et al. (1990) reports correlations of 0.07 between age and work performance and 0.18 between work experience and age. The results vary widely between industrial sector and job content. Avolio et al. used worker ratings by hierarchical superiors as the bases for their classifications of work performance. The problems associated with this method are reviewed in Section 3.4.

Performance levels of control, supervisory functions, and simple manual activities (Craft I) show correlations of 0.06 with age and 0.13 with work experience. In more complex assembly work, operations placing high demands on fine motor function and maintenance work (Craft II), the correlation between performance level and age is 0.13 and between performance level and work experience is 0.23 (Figure 3.11).

A French cross-section study embracing over 3.9 million workers in 70,000 companies shows that workers aged over 40 are approximately 5% more productive than those in the 35 to 39 age group. Workers under age 30 are between 15 and 20% less productive (Aubert and Crépon 2003).

The figures published by Aubert and Crépon are based on linked employer–employee data collected by the French fiscal authorities and integrated into economic production functions by the general method of moments (GMM). Allowance was made only for certain intervening and confounding variables.

The study by Goebel and Zwick (2009) of a random sample of 8571 subjects also used the GMM procedure, but applied various corrections, especially in regard to dynamic effects.

Figure 3.12 shows the age sequences in the studies by Aubert and Crépon (2003) and Goebel and Zwick (2009). It should be noted that

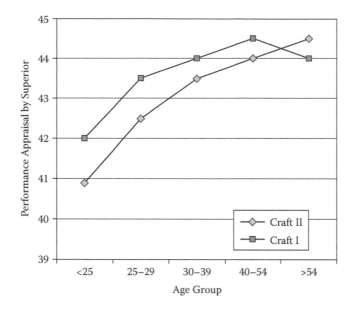

Figure 3.11 Performance levels for two different industrial production job requirements. (*Source:* Avolio, B.J., Waldmann, D.A., and McDaniel, M.A. 1990. *Academy of Management Journal*, 33, 407–422. With permission.)

a comparison of the figures presented in the two publications is not recommended. Only the curves in the graph are of interest.

Analyses showed that younger workers in the 30 to 34 age group performed worse (with the exception of the spike in the Goebel and Zwick curve) than those in the age 35 to 39 reference group, and all the older age groups performed better. Although both studies show a fall in productivity in the 55 to 59 age group, they did not fall below the level of the reference group. The results for the 55 to 59 age group were, however, influenced by a number of selection effects and should therefore be interpreted with caution.

A publication by Boersch-Supan and Weiss (2007) interprets the work quality of teams as a measure of productivity. The study relates to assembly line production at a truck manufacturing company and shows a steady rise in quality performance of team members from age 20 onward. It is impossible for obvious reasons to separate the effects of worker age and work experience in this study. The authors conclude that although older workers' error rates were slightly higher, their superior experience helped them avoid serious quality defects. Other published data are available, for example, from Norway (Haegeland and Klette 1999) and the United States (Haltiwanger et al. 1999).

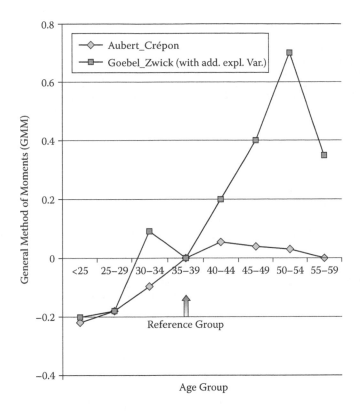

Figure 3.12 Corporate productivity calculated by GMM for French and German study populations. (*Sources:* Aubert, P. and B. Crépon. 2003. *Economic et Statistique*, 363: 95–119; Goebel, C. and T. Zwick. 2009. Centre for European Economic Research, Discussion Paper 09-020, Mannheim. With permission.)

Econometric cross-section studies can contain a significant number of sources of error, including the need to estimate age group size in a company's workforce, inclusion of part-time workers, the degree to which capital productivity influences productivity (e.g., older workers frequently work with old equipment), staff age structure, company age, and other issues. Such studies often tend to ignore dynamic effects, for example, production output in period i depends partly on the output in period $i-1$. Refer to Boersch-Supan and Weiss (2007) or Ours and Stoeldraijer (2010) for further results of econometric studies on worker age.

3.8   Opportunities for improving productivity

There are two ways to react constructively to discrepancies between capabilities of older workers and job demands. The first calls for ergonomic

Table 3.1 Recommendations for Job Design

Workplace Design Recommendations for Reducing Stress	Reasons
Eliminate prolonged forced postures	Forced postures involve static muscular effort. Older workers are less able to stabilize their bodies in unfavorable postures for long periods.
Design for regular posture changes	Successive changes of posture allow recovery of posture muscles.
Eliminate or limit lifting and carrying heavy loads, pulling and pushing	Older workers have less muscular power and often suffer from joint disorders. This reduces their load-handling capability and increases accident probability.
Reduce specified work speed	Older workers have poorer neuromuscular coordination and cannot adjust as quickly to rapid changes in work content.
Reduce static holding work	Older workers have less muscular power and often suffer from joint disorders. This reduces their ability to perform static holding work.
Reduce need for high energy turnovers	Cardiovascular function deteriorates with advancing age.
Make allowance for greater body weights of older workers	Greater body weight creates a need for more space mobility, especially in cramped conditions.
Reduce noise stress	Older workers frequently have hearing impairments and tend to suffer more acutely from loud noise.
Reduce heat stress	Older workers are less able to cope with physical stress in excessively warm ambient conditions.
Increase illumination and reduce dazzle	Older workers need better illumination to compensate for deterioration in visual powers.
Avoid shift work	Shift work has adverse effects on most metabolic diseases and can easily exacerbate symptoms in older workers suffering from (recurrent) gastrointestinal tract disorders or diabetes, for example. Older persons have more difficulty adjusting to irregularities in sleep–wake rhythm caused by shift work and have serious problems adjusting to switches from day shifts to night shifts.

(continued)

Table 3.1 Recommendations for Job Design (continued)

Workplace Design Recommendations for Reducing Stress	Reasons
Eliminate risks of falling accidents	Although there is no direct association between worker age and accident risk, older workers are at greater risk of falling. This risk should be reduced by eliminating cables, hoses, and piping lying on workplace floors and correcting uneven floor surfaces. Older workers, especially women, suffering from osteoporosis are at greater risk of bone fractures if they fall.

job design and organization and the second for worker training. The first approach is preferable because it offers social advantages for workers and financial advantages for employers because employers still have the well-designed workplaces after the workers retire. The second alternative can be regarded as a sensible complementary measure.

Table 3.1 summarizes the most important considerations for designing workplaces for older people. It is obvious that the improvements would be equally beneficial for younger workers. It is difficult to identify design features that are preferable *a priori* for older workers because all workers can benefit from such features. Consequently, the ideal approach is to design for all. The list should not be seen as a deficit model similar to the design recommendations proposed by the World Health Organization in the 1990s (WHO 1993). The evidence discussed above shows that it is incorrect to talk of deficits of older workers in relation to job demands and declining work productivity. Consequently, the focus of action to help older workers in performance of their jobs should be directed at work organization and behavioral ergonomics.

3.9   Conclusions

Ergonomic workplace design, optimal work organization, and health coaching of employees, when modified as necessary to allow for differing needs and abilities of various age groups, are justified demands in both micro- and macro-economic terms. However, cost limitations imposed on new technical and technological developments and the associated organizational changes frequently create conflicts of interest in attainment of the objectives.

It is therefore not surprising that many unfavorable stress situations and the ergonomic and the organizational design weaknesses that create them are discovered post hoc as results of productivity fluctuation and absenteeism for sickness. This is why, in the authors' experience, prospective or even good strategic workplace design in industry is still the exception rather than the rule. The reasons for this deplorable state are numerous and include the lack of knowledge of ergonomics and organizational skills designers, production planners, procurement officers, and staffing planners. This knowledge deficit can easily lead to work systems that are ergonomically and organizationally suboptimal, especially in cases of differing opinions on age-related needs and deficiencies.

It is, however, necessary to differentiate between large and smaller companies. Many large corporations have invested in resources that allow them to pursue enlightened policies on work safety and health risk minimization. Some have introduced strategic age management programs. Many smaller companies, especially those manufacturing and supplying components to large corporations, may lack the resources and the specialized knowledge of modern safety and health risk management designed for the needs of older workers.

References

Adenauer, S. 2002. Die Älteren und ihre Stärken: Unternehmen handeln. (Older workers and their strengths: Industry takes action). *Angewandte Arbeitswissenschaft*, 177, 36–52.

Aubert, P. and Crépon, B. 2003. Are older workers less productive? *Economie et Statistique* 363, 95–119.

Avolio, B. J., Waldmann, D.A. and McDaniel, M.A. 1990. Age and work performance in non managerial jobs: The effects of experience and occupational type. *The Academy of Management Journal* 33(2), 407-422.

BLS (Bureau of Labor Statistics). 1997. Industry productivity measures. In *Handbook of Methods*, BLS Bulletin 2490. Washington, pp. 103–109.

Boersch-Supan, A. and Weiss, M. 2007. Productivity and the age composition of work teams: Evidence from the assembly line. Mannheim Research Institute for the Economics of Aging. Report 148.

European Standard EN 614-1. 2006. Safety of Machinery: Ergonomic Design Principles. Brussels: European Union.

Fricke, R. 1961. *Die Grundlage der Produktivitätstheorie (Productivity Theory)*. Frankfurt am Main: Klostermann.

Fuchs, T. 2010. Potentiale des DGB Index Gute Arbeit für die betriebliche Anwendung und arbeitswissenschaftliche Forschung. Replik auf den Artikel von G. Richenhagen und J. Prümper in der ZfA 2/2009. (Potential for use of the DGB Index of Good Jobs in industry and ergonomic research. A reply to the article of G. Richenhagen and J. Prümper in the ZfA 2/2009). *Zeitschrift für Arbeitswissenschaft*, 1, 3–15.

Gelderblom, A. 2006. The relationship of age with productivity and wages. In: European Commission: *Ageing and employment: Identification of good practice to increase job opportunities and maintain older workers in employment*, pp. 67–84, 216–228. Brussels: European Commission.

German Journal of Ergonomics (Zeitschrift für Arbeitswissenschaft) 2010, 1. Special issue.

Goebel, C. and Zwick, T. 2009. Age and productivity: evidence from linked employer–employee data, Discussion Paper 09-020. Mannheim: Centre for European Economic Research.

Haegeland, T. and Klette, T.J. 1999. Do higher wages reflect higher productivity? In Haltwanger, J. et al., Eds., *The Creation and Analysis of Employer–Employee Matched Data*. Amsterdam: North Holland.

Haltiwanger, J.C., Lane, J.I., and Spletzer, J.R. 1999. Productivity differences across employers: The roles of employer size, age, human capital. *American Economic Review*, 89, 94–98.

Ilmarinen, J., Tuomi, K., and Klockars, M. 1997. Changes in the work ability of active employees over an 11-year period. *Scandinavian Journal of Work Environmental Health*, 23, 49–57.

International Institute for Empiric Social Economy (INIFES). 2010. TNS Infratest Sozialforschung GmbH, Munich.

International Institute for Empiric Social Economy (INIFES). 2008. Arbeitsbedingungen und Arbeitsfähigkeit bis zur Rente. (Working conditions and work capability in the run-up to pension age). Ergebnisse aus der Erhebung zum DGB Index Gute Arbeit.

Isernhagen, S. 1992. Functional capacity evaluation: rationale, procedure, utility of the kinesiophysical approach. *Journal of Occupational Rehabilitation, 2,* 157–168.

Karhu, O., Kansi, P., and Kuorinka, I. 1977. Correcting working postures in industry: a practical method for analysis. *Applied Ergonomics, 8,* 199–201.

Landau, K., Meschke, H., Brauchler, R. et al. 2007. Ergonomics diagnosis components in rehabilitation: statistical evaluation of an assessment instrument. *Ergonomics* 50, 1871–1896.

Landau, K., Rademacher, H., Meschke, H. et al. 2008. Musculoskeletal disorders in assembly jobs in the automotive industry with special reference to age management aspects. *International Journal of Industrial Ergonomics*, 38, 561–576.

Marchetti, C. 2002. Productivity versus age. International Institute for Applied Systems Analysis, Laxenburg, Austria, Contract 00-155.

OECD (Organisation for Economic Cooperation and Development). 2006. Employment Outlook. http://www.oecd.org/els/employmentoutlook/EmO2006

OECD (Organisation for Economic Cooperation and Development) 1998. *Society at a Glance: OECD Social Indicators*. Paris.

Ours, J. C. van and Stoeldraijer, L. 2010. Age, wage, and productivity, Discussion Paper 4765. Bonn: Institute for the Study of Labor.

Puch, K. 2009. Erwerbsbeteiligung älterer Arbeitnehmer (Involvement of older workers in gainful employment). http://www.destatis.de/jetspeed/portal/cms/Sites/destatis/Internet/DE/Content/Publikationen/STATmagazin/Arbeitsmarkt/2009__01/2009__01Erwerbsbeteiligung,templateId=renderPrint.psml

REFA Fachausschuss Chemie. 1987. *Handhaben von Lasten*. Darmstadt: Verband für Arbeitsstudien e.V.

Sanders, M. 2009. Job design factors in the workplace that support successful aging for older workers. Ph.D. Dissertation, Walden University, College of Social and Behavioral Studies.

Schaub, K.H. 2004. Automotive Assembly Worksheet (AAWS). In Landau, K., Ed., *Montageprozesse gestalten*. Stuttgart: Ergonomia, pp. 91–111.

Schaub, K.H., Helbig, R., and Spelten, C. 1999. IAD-BKA: eine ergonomische Arbeitsplatzanalyse als Voraussetzung der Arbeitsgestaltung in Klein- und Mittelbetrieben zur Vermeidung arbeitsbedingter Erkrankungen. Darmstadt.

Skirrbek, V. 2003. Age and individual productivity: A literature survey. Working Paper 2003-028. Rostock: Max Planck Institute for Demographic Research.

Walker, B.F. 2000. The prevalence of low back pain: A systematic review of the literature from 1966 to 1998. *Journal of Spinal Disorders & Techniques* 13, 205–217.

Waters, T.R., Putz-Anderson, V., and Garg, A. 1994. *Applications Manual for the Revised NIOSH Lifting Equation*, Publication 94-110. Cincinnati: NIOSH).

World Health Organization. 1993. *Ageing and Working Capacity*. Technical Series 835. Geneva.

chapter four

Human factors related to success of total productive maintenance

Jorge Luis García Alcaraz, Alejandro Alvarado Iniesta, and Aidé Aracely Maldonado Macías

Contents

4.1   Introduction

Total productive maintenance (TPM) is one of the main tools to achieve efficiency and competitiveness by enhancing quality and reducing time and production costs. TPM is generally executed jointly with a total quality management (TQM) program based on a continuous effort to improve processes and production resources (Wikoff 2007). TPM is highly effective in enterprises that engage in automatic and sequential activities through intensive use of machinery. TPM does not require huge amounts of economic investment to achieve the exploitation of existing installations because it is based on planning (Cooke 2000, Wikoff 2007).

However, the quality of the finished product depends on the machine operators, the supervisors, the decision makers, and the enterprise's material and equipment suppliers. The traditional models measure efficiency through machinery indices that also indirectly measure worker efficiency.

TPM is not a new idea; it is simply the next step in the evolution of good maintenance practices (Dinesh and Triphati 2006, Wikoff 2007). TPM and predictive maintenance (PM) are sometimes confused. PM focuses on scheduled maintenance. TPM involves analysis of life cycles of products and equipment to reduce the incidence of failures, production defects, and accidents. It is an aggressive strategy that concentrates on improving production design and function (Chandra and Shastri 1998, Wikoff 2007). TPM also involves all levels on workers in an organization who play roles in production activities.

TPM attempts to increase the availability and effectiveness of equipment by maintaining it in optimal service condition to increase its life cycle. As a result, the objective of improving quality and efficiency is achieved with a minimum investment (Cooke 2000). The concept involves controlling and decreasing variation in the production process (Reed 1996).

However, when TPM is not properly implemented, six losses divided into three categories can occur (Tajiri and Gotoh 1992, Sachdeva et al. 2008): (1) production lines are idle or must stop completely; (2) operations are executed at speeds far below the capability of the equipment, and (3) defects in the product or flaws in equipment operation increase. The six losses are:

1. Frequent machinery failures
2. Time wasted for preparation and set-up
3. "Microstops" (short production interruptions that add up to a lot of time wasted)
4. Slower production speed and frequent bottlenecks
5. Reduced quality of final product
6. Restart time

Conversely, adequate implementation of TPM produces several benefits:

1. Maintenance activities are planned and controlled (Tajiri and Gotoh 1992, Eti et al. 2004) allowing reduction of maintenance workforce.
2. Reduction of indirect workforce (Takahashi and Osada 1989, Chan et al. 2005).
3. Improved relations among operators and generation of ideas that contribute to enterprise efficiency (Chand and Shirvani 2000, Cua et al. 2001, Gosavi 2006, McKone et al. 2001, Chan et al. 2005, Aichlmayr 2009).
4. Increased product quality and customer satisfaction (Ashayeri 2007).
5. More efficient scheduling based on better equipment reliability and product standardization ((Alsyouf 2009, Dowlatshahi 2008).

Case studies report results of application of TMP in various enterprises For example, McKone et al. (1999) reported an analysis of U.S. industries that implemented TPM and related its benefits. Eti et al. (2004) reported the implementation and benefits derived from TPM. Sachdeva et al. (2008) reported optimized results in a plant dedicated to paper production. Dowlatshahi (2008) made an empirical analysis of the role of TPM in the manufacturing industry. Pinjala et al. (2006) analyzed the importance of TPM and concluded that it must be part of the strategy of an enterprise that wishes to be competitive. Vinodh (2010) noted that an agile and sustainable production system can be achieved by applying TPM. He based his work on reports from Cua et al. (2001) who related the benefits of TPM in combination with TQM and just-in-time (JIT) programs.

A lot of benefits result from implementation of TPM. However, the activities of employees required to achieve these benefits in industry remain unclear. Specifically, what do workers, managers, supervisors, and machinery suppliers have to do to achieve the benefits? Unfortunately, many authors limited their studies to defining the TPM implementation processes in specific case studies (Eti et al. 2006, Pinjala et al. 2006, Zhou and Zhu 2008, Aissani et al. 2009). The authors of these papers were generally the managers or consultants responsible for TPM implementation in a single enterprise and they report isolated observations and facts in one place for a specified time.

A few studies concerned the key success factors of TPM, specifically, human factors such as responsibilities of managers, supervisors, operators, and suppliers of machinery and equipment. Therefore, the objective of this chapter is to present a descriptive and multivariable analysis in which the human key success factors of TPM associated with those groups can be identified. The intent is to examine the relationships and responsibilities of these groups in an effort to improve the adoption of TPM by an enterprise.

4.2   Methodology

The methodology was based on a survey to identify the types of activities performed by managers, supervisors, operators, and suppliers that are considered important to TPM success. After completion of the surveys, the data were collected and analyzed and conclusions drawn. The study was conducted in several phases described below.

4.2.1   First phase: Activity identification and survey development

First, we performed a literature review to identify the activities related to human factors that determine the success of TPM. We considered 42 references relevant (13 related to administration, 9 to operators, 12 to supervisors, 8 to suppliers). Five among the 42 concerned general benefits of implementation. The information obtained from the literature allowed us to develop a preliminary questionnaire that was structured and distributed to 88 managers or supervisors of TPM in enterprises located in Ciudad Juarez, Chihuahua, Mexico.

The original questionnaire included blank space for use by respondents to report other activities and the corresponding benefits they considered important. A total of two new administrative activities were identified, bringing the total activities in the final questionnaire to 44. The final questionnaire contained five sections:

Questionnaire Section	No. of Questions
Administration and management	15
Operators	9
Supervisors	12
Suppliers	8

Five questions spread throughout the five categories concerned the benefits of good TPM practices. The questionnaires were based on a Likert scale (Likert 1932) citing values between 1 and 9 (1 = no importance; 9 = extreme importance) to assess the influence of each activity on the success of the TPM program.

4.2.2   Second phase: Execution of questionnaire

A group of 769 managers and maintenance supervisors in Ciudad Juarez were contacted by telephone to obtain responses to the questionnaires. They were selected from a directory of enterprises and personnel listed by the Maquiladoras Association of maintenance professionals. Appointments were made to visit workplaces where possible. Some respondents preferred to respond electronically.

Because the questionnaires were comprehensive, one to three visits were arranged. Any response requiring more than three visits was abandoned. Similarly, the electronic responders received reminders after 3 weeks. After two additional reminders the subjects were considered nonresponders and the efforts abandoned.

4.2.3   Third phase: Data capture and questionnaire validation

Data were extracted from the questionnaire and entered into spreadsheets of SPSS 18 (Statistical Package for the Social Sciences software). However, before doing any analysis, the questionnaire was validated using Cronbach's Alpha Index (Cronbach, 1951). The rating exceeding 0.7 was considered valid.

4.2.4   Fourth phase: Descriptive analysis of data

The next step was descriptive analysis of the information. The median was taken as a central tendency measure because the data were numeric (Denneberg and Grabisch 2004, Pollandt and Wille 2005, Tastle and Wierman 2007). High median value indicated that an activity was of high importance for TPM; low values indicated lack of importance. Similarly, the first and third quartiles were measured. The difference between was designated the interquartile range (IR) that represented 50% of the data and included the median represented by the second quartile (Tastle and Wierman 2007).

High IR values indicated a lack of consensus among the respondents related to the importance levels of these activities. Low values represented a low level of dispersion and a higher consensus about activity importance among the responders.

4.2.5   Fifth phase: Exploratory factor analysis (EFA) by sections

The feasibility of the exploratory factor analysis (EFA) was determined and the correlation matrix was analyzed. Correlations among activities exceeded 0.3 (Nunnally 1978, Nunnally and Bernstein 2005). The diagonal of the anti-image matrix of correlation was analyzed to determine sample adequacy. Also, the Kaiser-Meyer-Olkin (KMO) index was calculated. Bartlett's sphericity test was applied to measure sample adequacy, and the commonalities of all the activities were analyzed to determine their contributions using 0.5 as a value limit (Lévy and Varela 2003).

To find the critical factors or latent variables in each of the four human factors that intervened in TPM success, a factor analysis by principal components method based on the correlation matrix was used. A factor was considered important if it yielded an eigenvalue of one or more. The search was limited to 100 iterations for the convergence of a result (Streiner and Norman 1995). With the objective of obtaining better interpretations of the critical factors, a rotation was done using the Varimax method (Lévy and Varela 2003). The activities that integrated the factors were identified by the high values of the factorial loads (Nunnally and Bernstein 2005).

4.3   Results

4.3.1   Sample composition

A total of 203 valid questionnaires from 72 enterprises located in several industrial parks in Ciudad Juarez, Mexico were returned; 68 came from managers and the remaining 135 were from supervisors. Figure 4.1 illustrates the types of enterprises and industrial sectors surveyed—the most common sectors were the automotive, electric, and electronic industries.

Figure 4.2 illustrates the professions of the subjects surveyed. Industrial engineering was the most common profession, followed by mechanical engineering, and electronic and mechatronic engineering.

4.3.2   Validating questionnaire, scale construction and identifying activities

Cronbach's alpha rates for the 44 initial activities were analyzed per section for each human factor relevant to TPM success. Table 4.1 illustrates

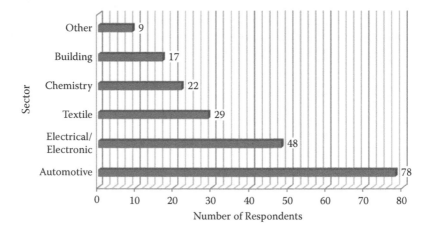

Figure 4.1 Sectors surveyed.

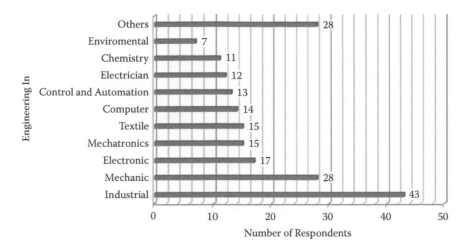

Figure 4.2 Careers surveyed.

Table 4.1 Questionnaire Validation

| | | Spearman-Brown | | |
Factor	Cronbach's Alpha	First Half	Second Half	Variables
Administration	0.861	0.804	0.816	15
Operators	0.834	0.795	0.797	9
Supervisors	0.843	0.798	0.809	12
Suppliers	0.832	0.853	0.795	8
Results	0.834	0.798	0.802	5

the validity results obtained with two tests (Cronbach's Alpha Index and the Spearman-Brown Index) and the number of variables for each section. All index values exceeded 0.7.

4.3.3   Descriptive analysis of sample

The descriptive analysis was done by sections according to the analyzed human factor. The next subsections discuss the main findings.

4.3.3.1   Information related to administration

Table 4.2 lists 15 activities that form the section of administrative activities associated with TPM success and displays the median, percentile 25, percentile 75, and the IR. Based on high median values, it can be said that to achieve satisfactory results from a TPM program, management must

Table 4.2 Descriptive Analysis: Management and Administration

Activities	Median	Percentiles 25	75	IR
Do managers and executives serve as examples of order and cleanliness?	7.05	4.72	8.49	3.77*
Are immediate supervisors and maintenance personnel committed to efficient machinery operation?	6.91	4.38	8.45	4.07
Is maintenance emphasized as a strategy to achieve program activities and quality?	6.9	4.54	8.47	3.93*
Do managers promote worker participation in maintenance and prolonging the life of equipment?	6.75	4.15	8.36	4.21
Can an operator request maintenance directly when a failure occurs?	6.5	3.22	8.49	5.27
Do top managers provide leadership of TPM programs?	6.48	4.17	8.27	4.1
Are the maintenance programs based on the useful lives of machines and their components?	6.35	4.16	7.98	3.82*
Is TPM progress tracked and evaluated?	6.22	3.62	7.99	4.37
Do people responsible for machine maintenance participate in informative meetings?	6.13	3.78	7.79	4.01
When are investments in equipment realized? Are maintenance requirements considered important criteria for acquisitions?	6.12	3.71	7.97	4.27
Is management involved in maintenance projects?	5.48	2.66	7.6	4.94
Are investments in innovative tools to facilitate maintenance returned?	5.27	2.89	7.31	4.42
Are returns on investments calculated for processes and equipment the enterprise patented?	4.59	2.01	7.29	5.28**
Are the workers motivated to express their opinions and ideas before maintenance decisions are made?	4.57	2.4	7.38	4.98
Are failures to meet TPM objectives explained?	4.05	1.8	6.44	4.65

* Activities with lowest interquartile ranges.
** Activities with highest interquartile ranges.

promote and practice neatness and cleanliness in work areas and coop-
erate with supervisors and workers responsible for machinery maintenance. These actions demonstrate TPM as a strategy to achieve quality
control. Management must work with team members and demonstrate
leadership in implementation of the program.

Low median values indicated activities that were less important for
TPM success (equipment monitoring, encouraging operators to give opinions and present ideas related to maintenance decisions, and explaining
the reasons for not meeting TPM objectives).

The three activities with the lowest interquartile ranges were management examples of order and cleanliness, maintenance programs, and
finally maintenance strategies for achieving quality. The activities are indicated by asterisks (*) in the descriptive analysis tables. Low values indicate
a consensus among respondents about the low values of the activities. It is
important to mention that two of these activities revealed the highest values in the median. This means that the low importance was corroborated.

Similarly, the item related to the availability of the equipment exhibited the highest IR values (marked ** in the table). The result means that
no consensus was reached about the value of an item. Thus, the item does
not contribute to a satisfactory TPM result and yielded one of the lowest
median values.

4.3.3.2   Information related to supervision

Table 4.3 lists the activities or items associated with human factors of
supervision. The data are ranked in descending order of the median
value. To achieve TPMs success, supervisors must keep work and maintenance records for all equipment, communicate with management about
the execution of maintenance programs, and use graphics to demonstrate
the status of equipment and components. These activities relate to control,
leadership, and communication of the results of TPM programs.

In connection with activities that showed lower median values, the
respondents considered that it was not important to communicate information about maintenance records to operators. These values indicated
that operator opinions were of low importance for TPM success.

Activities that demonstrated high median values yielded low IR values, indicating that TPM program leadership, graphics for sharing information, and maintenance records were important. Two activities that had
low median values had high IR values. This means that the respondents
did not achieve a consensus about the real values of the two items.

4.3.3.3   Information related to operators

Table 4.4 illustrates the descriptive analysis of the human factors related
to operators; medians are ranked in descending order. The highest values are represented by the emphasis of the enterprise on orderly use and

Table 4.3 Descriptive Analysis: Supervision

Activities	Median	Percentiles 25	75	IR
Is a maintenance record kept for each machine?	7.6	5.39	8.81	3.42*
Does leadership promote TPM programs on behalf of the maintenance group?	7.13	5.26	8.5	3.24*
Do graphics demonstrate whether the manufacturing process is in control?	7.05	4.62	8.6	3.98*
Is maintenance programming based on a consensus with production staff?	6.83	4.34	8.44	4.1
Does production and engineering leadership participate in the execution of TPM programs?	6.79	4.1	8.38	4.28
Do employees work actively to improve existing machinery?	6.4	3.65	8.03	4.38
Are the causes of malfunctions identified, recorded, and developed into statistics?	6.37	4.04	8.26	4.22
Is part of every workday dedicated exclusively to maintenance?	6.1	3.6	8.08	4.48**
Does the maintenance department support operators' needs related to preventive maintenance on machinery?	5.37	3.47	7.91	4.45
Does the maintenance department inform operators of completed maintenance activities?	5.31	3.11	7.85	4.74**
Are operator observations included in maintenance records?	3.36	1.51	6.17	4.66**
Are operator opinions considered in programming routine maintenance?	3.21	1.45	5.65	4.19

* Activities with lowest interquartile ranges.
** Activities with highest interquartile ranges.

storage of tools and accessories. Cleanliness inspections of the work areas and warehouse are conducted regularly. The low values of the medians of the factors are associated with the constant movements of spare parts in the warehouse area, the critical systems that may cause machines to fail, and the levels of knowledge of operators about maintenance programs for the equipment. The last items concern equipment information that should be communicated to operators.

Table 4.4 Descriptive Analysis: Operators

Activities	Median	Percentiles		IR
		25	75	
Does the enterprise emphasize workplace cleanliness and order?	7.54	4.86	8.87	4.01**
Is cleanliness maintained in the warehouse area?	7.23	5.55	8.51	2.96*
Are work areas clean and orderly?	6.74	4.8	8.16	3.35*
Are workers responsible for start-up maintenance of the machines they operate?	6.72	4.64	8.26	3.63
Do operators seek training in relation to maintenance of their machines?	6.61	4.7	8.42	3.72
Are all areas of the enterprise always clean?	6.54	4.88	8.29	3.42*
Are constant changes in warehouse operations noted and communicated?	6.42	3.8	8.22	4.42**
Do operators know about critical systems that can lead to machine failure?	6.24	4.19	7.78	3.58
Do operators know about maintenance programs for their equipment?	5.1	2.65	7.26	4.61**

* Activities with lowest interquartile ranges.
** Activities with highest interquartile ranges.

The lower IR values are associated with two of the activities that displayed the highest medians. We can conclude that activities related to cleanliness are highly important for TPM success. Similarly, the activities or items with high IR values concern operator knowledge about critical systems that can cause machine failure and about equipment maintenance programs.

4.3.3.4   Information related to suppliers

Table 4.5 illustrates the descriptive analysis of the supplier activities considered vital for TPM success. The activities with the highest medians concern adequate manuals for machine maintenance, compliance with equipment warranties, and technical support for maintenance activities. Warranty issues are important because the enterprise will be responsible for machines after their warranties expire.

Similarly, the activities with the lowest median values refer to mutual training programs for equipment suppliers and buyers, in-plant training,

Table 4.5 Descriptive Analysis: Suppliers

| Activities | Median | Percentiles | | IR |
		25	75	
Do suppliers furnish adequate manuals for machine maintenance?	7.2	5.13	8.57	3.45*
Do suppliers fulfill equipment warranties?	7.16	5.3	8.55	3.25*
Do suppliers provide technical support for equipment maintenance?	7.14	4.77	8.55	3.78*
Are suppliers responsible for installing and adjusting machinery they sell?	6.23	3.57	8.3	4.73**
Does the enterprise execute maintenance agreements with machine suppliers?	6.08	3.71	7.64	3.93
Do buyers and suppliers participate in mutual training programs?	5.36	3.11	7.52	4.41**
Do machinery suppliers provide in-plant training for the buyer's maintenance staff?	5.09	3.11	7.4	4.29
Do machine suppliers conduct training at their own facilities for the buyer's maintenance staff?	4.89	2.64	7.19	4.55**

* Activities with lowest interquartile ranges.
** Activities with highest interquartile ranges.

and training on supplier premises. Note that training by suppliers is not rated as important.

The column showing IRs indicate that activities that have minor values (one asterisk) are the same ones that achieved the highest median values—indicating their high importance. The high values of the IR that indicate no consensus about values are associated with two activities that have low median values and concern supplier-provided training.

4.3.3.5   *Information related to results*

Table 4.6 demonstrates the results of the descriptive analysis of the four human factors and their specific activities. The main result was that the management created and communicated a vision focused on quality and maintenance. The next result is that the management demonstrates leadership in the execution of TPM programs. The production equipment and machinery of the enterprise present a competitive advantage in the market. However, it is important to mention that in reality the first four factor

Table 4.6 Descriptive Analysis: Action Results

		Percentiles		
Activities	Median	25	75	IR
Management creates and communicates a vision centered on quality and maintenance.	6.656	4.403	8.151	3.748*
Management promotes the execution of TPM programs.	6.51	3.446	8.159	4.713
Enterprise equipment helps achieve a competitive advantage.	6.368	4.462	8.026	3.564*
All department heads accept their responsibilities for TPM.	6.293	3.409	8.231	4.822
Users play an important part in maintenance decisions.	4.912	2.269	7.585	5.316**

* Activities with lowest interquartile ranges.
** Activities with highest interquartile ranges.

medians between 6 and 7 represent values differing by hundredths and are thus relatively similar in importance.

Similarly, the low median set for the equipment user's role in making decisions about maintenance indicates high importance attributed to participation in deciding maintenance issues. Minor IR values are related to results that achieved the highest values in the median and the IR values corroborate the values of the variables. The highest IR values indicate activities with the lowest median values for integrating users in decision-making processes.

4.3.4   Factor analysis

The four human factors considered vital for TPM success were analyzed. Table 4.7 lists them (managers, supervisors, operators, and machinery suppliers). The KMO achieved were in all cases greater than the 0.8 value recommended as the inferior limit. In addition, the values of the chi square, degrees of freedom (DF), and significance levels that correspond to those factors are less than 0.05, indicating significance.

Also, the determinant of the correlation matrix is small and indicate that the application of factor analysis was feasible. The table shows the human factors analyzed and the variance percentages. For example, the factors associated with administration consist of three components that explain the 54.67% of the total of the variance represented by the data.

Table 4.7 Reliability of Factor Analysis

Factor	KMO	Bartlett Sphericity Test			Determinant	Number of Factors	% Explained Variance
		Chi Square	DF	Significance			
Administration	0.876	1084.33	105	0.00	0.004	3	54.67
Operators	0.835	670.74	36	0.00	0.034	2	56.92
Supervisors	0.813	810.09	66	0.00	0.016	3	58.92
Suppliers	0.852	713.15	28	0.00	0.028	2	62.18

Table 4.8 Factor Analysis: Administration

Activities	Response	Factor, Variance
Do supervisors and maintenance personnel compromise on issues of machine functionality?	0.756	Leadership (26.37%)
Do managers participate in executing TPM programs?	0.718	
Do enterprise directives promote the participation of workers in maintenance and equipment conservation?	0.716	
Is management involved in maintenance projects?	0.601	
Are maintenance programs monitored; Is there progress monitored?	0.597	
Do managers promote cleanliness and order in work areas?	0.583	
Are innovative tools purchased in an effort to simplify maintenance?	0.79	Investments (16.37%)
Are maintenance requirements considered when equipment is purchased?	0.784	
Are failures to meet TPM objectives explained?	0.734	Communication (11.92%)
Are returns on investments calculated for processes and equipment the enterprise patented?	0.579	

4.3.4.1 Factor analysis: Management and administration

Table 4.8 illustrates the factorial loads associated with management. The main factor concerns management leadership of TPM programs and explains the 26.37% variance. The second factor concerns investments in new and flexible tools to simplify equipment maintenance. Maintenance requirements should be considered key factors when an enterprise invests in machinery. This explains the 16.37% which, when added to the last factor (requirement for administration authorization of purchases) accounts for 42.74%. This factor is grouped with administration because administration is responsible for authorizing new equipment purchases.

A third factor is communication among the managers of the workers responsible for maintenance. Communication is vital because management controls resource availability and finances. The operators understand the operative needs of the equipment. This factor explains the

Table 4.9 Factor Analysis: Supervision

Activities	Response	Factor, Variance
Does the maintenance department support operators in achieving preventive maintenance?	0.694	Orientation and communication (24%)
Is part of every workday dedicated exclusively to maintenance?	0.668	
Does the maintenance department inform operators of completed maintenance activities?	0.652	
Are graphics used to determine whether manufacturing processes are controlled?	0.627	
Do production and engineering leaders aid in executing TPM programs?	0.607	
Are records identifying machinery failures and malfunctions kept and developed into statistics?	0.572	
Is there leadership in executing TPM programs on behalf of maintenance?	0.554	
Does maintenance planning consider operator opinions?	0.867	Operator integration (18.25%)
Are operator observations included in maintenance records?	0.815	
Is maintenance planned according to a consensus with the production department?	0.71	Coordination and continuous improvement (16.67%)
Are daily records of work and maintenance kept for all machines?	0.706	
Are attempts made to achieve continuous improvement of existing machinery?	0.578	

11.92%. When added to the previous factors, management issues total 54.67% of the total variability of the data.

4.3.4.2   Factor analysis: Supervision

Three subfactors explain this human factor and the factorial loads are indicated in Table 4.9. The first covers orientation, communication, and support of the TPM department to supervisory activities related to leadership and good practices to obtain satisfactory results. This factor explains 24% of the variance. The second factor of TPM supervision concerns integration of operators in the decision-making process, enterprise maintenance management, and maintenance of daily records. This factor explains 18.25% of the variability of all the analyzed items. The total for the first and second factors is 42.25%.

The third human factor associated to supervision and related to the TPM deals with coordination between the maintenance and production departments to overcome their normally antagonistic positions concerning equipment availability. Production wants all equipment to meet production orders; maintenance focuses on required corrective measures to keep machines running. This factor covers integrated activities leading to continuous improvement of production processes and daily recording of maintenance activities and machine malfunctions. This factor explains the 16.67% of the variability of the analyzed data and with the two previous factors accounts for 58.92%.

4.3.4.3 Factor analysis: Operators

This human factor consists of two main components described in Table 4.10. The first factor is associated with order and cleanliness in operator work areas, specifically training to use machinery properly, measures to improve work flow, and keeping machinery clean and ready to work. The factor accounts for 34.1% of the variance of all the analyzed data. The second component is worker knowledge of the equipment so that he or

Table 4.10 Factor Analysis: Operators

Activities	Response	Factor, Variance
Are work areas clean and orderly?	0.888	5S and training (34.10%)
Are all areas of the enterprise always clean?	0.824	
Do supervisors emphasize the need to store tools and materials properly?	0.761	
Do operators seek training in maintenance?	0.625	
Are warehouse areas always clean?	0.589	
Do operators know about critical machine systems that may fail?	0.769	Knowledge administration (22.83%)
Are changes in warehouse areas communicated?	0.625	
Do operators know about maintenance programs for their equipment?	0.616	
Are operators responsible for start-up maintenance of their equipment?	0.606	

Table 4.11 Factor Analysis: Suppliers

Activities	Response	Factor, Variance
Do suppliers offer technical support for maintenance of the equipment they sell?	0.828	Technical support and warranties (36.09%)
Do suppliers provide adequate maintenance manuals for machinery they sell?	0.791	
Does the enterprise execute maintenance agreements with suppliers of machinery?	0.773	
Do suppliers comply with their warranties?	0.767	
Do suppliers provide training on their premises for the enterprise's maintenance staff?	0.844	Training (30.09%)
Do suppliers provide training for the enterprise maintenance staff on enterprise premises?	0.841	
Are programs in place to provide mutual training of relevant supplier and enterprise employees?	0.785	

she can handle simple maintenance if necessary. The total for the operator factor is 22.83% of the variance; the total for both factors is 56.93%.

4.3.4.4   Factor analysis: Suppliers

This human factor has the two components indicated in Table 4.11. The first explains 36.09% of the total variance and refers to the technical support provided to an enterprise by machinery supplies, the provision of equipment manuals and technical instructions, the suppliers' compliance with warranties after equipment is installed. The second component is training of the enterprise's employees in the use and maintenance of purchased equipment either at the enterprise's premises or at suppliers' facilities. This factor explains the 30.089% of the variability; both factors total 66.18%.

4.3.4.5   Analysis of all factors

This category covers a total of 10 factors that explain the 65.4% total variance of the data. In addition to factors and variances, the right-most column lists responsible groups (Table 4.12). For example, the first factor is focus on maintenance and continuous improvement. The objective is to find root causes of problems related to equipment deficiencies and emphasize continuous improvement programs to increase efficiency. These

Table 4.12 All-Factor Analysis

Activities	Response	Factor, Responsibility, Variance
Have causes of machine failures and malfunctions been identified and are statistics recorded and maintained?	0.704	Maintenance and continuous improvement focus: Management (10.61%)
Are workers actively involved in improving existing machinery?	0.66	
Is progress of maintenance programs monitored and evaluated?	0.59	
Are workers responsible for machine maintenance involved in relevant meetings?	0.583	
Are the maintenance programs based on the useful life of a system and components of the machines?	0.555	
Is management personally involved in maintenance projects?	0.517	
Do the suppliers fulfill equipment warranties?	0.695	Warranties and supplier relations: Suppliers (8.74%)
Do suppliers provide technical support for equipment maintenance?	0.687	
Does the enterprise execute maintenance agreements with equipment suppliers?	0.678	
Are suppliers responsible for installing and adjusting purchased machinery?	0.619	
Do suppliers provide adequate maintenance manuals for their equipment?	0.545	
Are production areas clean and orderly?	0.875	5S: Operators (8.32%)
Are all areas of the enterprise clean during all working hours?	0.784	
Does the enterprise emphasize the need to store tools and accessories appropriately?	0.679	
Do operators seek training in maintenance?	0.509	
Are operators questioned about redactions of maintenance records?	0.727	Operator integration: Supervisors (7.77%)
Do maintenance programs account for operators' opinions?	0.724	

(continued)

Table 4.12 All-Factor Analysis (continued)

Activities	Response	Factor, Responsibility, Variance
Are workers motivated to express their opinions and ideas before maintenance decisions are made?	0.713	
Does the maintenance department inform operators of maintenance to be performed on their machines?	0.603	
Does the maintenance department support operators in their efforts to foster preventive maintenance?	0.52	
Do suppliers provide in-plant training for the enterprise's maintenance staff?	0.793	Training: Suppliers (3.32%)
Do suppliers provide training for their maintenance staffs at the enterprise's plant?	0.71	
Are programs in place for mutual training of buyers and suppliers?	0.574	
Is part of a day dedicated exclusively to maintenance?	0.685	TPM Planning: Supervisors and management (6.08%)
Is maintenance emphasized as a strategy to achieve quality and activities programming?	0.64	
Do enterprise directives promote worker participation in maintenance and equipment conservation?	0.666	TPM culture: Supervisors and management (5.96%)
Are critical systems of machines where failures can occur known?	0.659	
Can an operator request maintenance directly when a failure occurs?	0.523	
When are investments in equipment or machinery realized? Is maintenance an important criterion in considering acquisitions?	0.724	Investments: Management (4.57%)
Are investments in innovative tools intended to facilitate maintenance?	0.59	
Can failure to achieve TPM objectives be explained?	0.768	TPM Control: Supervisors (3.48%)
Was the maintenance program determined by consensus with the production department?	0.647	
Is a daily record of work and maintenance kept for each machine?	0.623	

factors explain 10.61% of the total variance of the data and are responsibilities of management.

The second factor concerns equipment suppliers and their relationships with the enterprise. Supplier responsibilities include provision of operation manuals and technical support, complying with warranties if problems arise, and possibly training. This factor explains 8.74% of the variability. It and the previous factor explain 19.35% of the variability. This interpretation also applies to the other eight factors cited in the table.

At this point it is necessary to explain more about the human factors that contribute in large part to TPM success. Operators, maintenance staff, and their supervisors have the most interactions with machinery and equipment. Administration and management are involved in five of ten factors, indicating a high level of involvement in the success of a TPM program. These groups make high-level maintenance decisions, support continuous improvement efforts, and plan and implement TPM projects and goals. They are also responsible for communicating results and authorizing investments in tools and equipment that can improve the maintenance of machinery. The operators' roles appear only in the third factor (order and cleanliness of work areas). Suppliers are cited in two factors and supervisors in three.

4.4   Conclusions

While TPM programs are specifically designed to improve production through effective equipment maintenance, these programs ultimately depend on people. That is why human factors are so important in TPM implementation. Humans (managers, supervisors, operators, and suppliers) perform the activities that guarantee program success.

Once each human factor was subjected to univariable analysis, we concluded that all these factors are essential for effective implementation of TPM in an enterprise. Leadership, investments in innovative equipment, close attention and maintaining records of equipment failures, good relations with suppliers, and maintaining orderly and clean work areas are all components of a well-run TPM program.

We can conclude, based on the multivariable analysis results and calculations, that management is an essential part of an effective TPM program even when it is not directly involved in machine maintenance and malfunction. This is clearly shown in management's involvement in five of the ten factors analyzed.

References

Aichlmayr, M. 2009. TPM: Healthcare for equipment. *Material Handling Management*, 64, 18–20.

Aissani, N., Beldjilali, B., and Trentesaux, D. 2009. Dynamic scheduling of maintenance tasks in the petroleum industry: A reinforcement approach. *Engineering Applications of Artificial Intelligence*, 22, 1089–1103.

Alsyouf, I. 2009. Maintenance practices in Swedish industries: Survey results. *International Journal of Production Economics*, 121, 212–223.

Ashayeri, J. 2007. Development of computer-aided maintenance resources planning (CAMRP): A case of multiple CNC machining centers. *Robotics and Computer Integrated Manufacturing*, 23, 614–623.

Chan, F.T.S., La, H.C.W., Ip, R.W. et al. 2005. Implementation of total productive maintenance: A case study. *International Journal of Production Economics*, 95, 71–94.

Chand, G. and Shirvani, B. 2000. Implementation of TPM in cellular manufacture. *Journal of Materials Processing Technology*, 103, 149–154.

Chandra, P. and Shastri, T. 1998. Competitiveness of Indian manufacturing: Findings of the 1997 manufacturing futures survey. *Vikalpa*, 23, 25–36.

Cooke, F.L. 2000. Implementing TPM in plant maintenance: Some organizational barriers. *International Journal of Quality and Reliability Management*, 17, 1003–1016.

Cronbach, L.J. 1951. Coefficient alpha and the internal structure of tests. *Psychometrika*, 16, 297–334.

Cua, K.O., McKone, K.E., and Schroeder, R.G. 2001. Relationships between implementation of TQM, JIT, and TPM and manufacturing performance. *Journal of Operations Management*, 19, 675–694.

Denneberg, D. and Grabisch, M. 2004. Measure and integral with purely ordinal scales. *Journal of Mathematical Psychology*, 48, 15–22.

Dinesh, S. and Tripathi, D. 2006. Critical study of TQM and TPM approaches on business performance of Indian manufacturing. *Total Quality Management and Business Excellence*, 17, 811–824.

Dowlatshahi, S. 2008. The role of industrial maintenance in the maquiladora industry: An empirical analysis. *International Journal of Production Economics*, 114, 298–307.

Eti, M.C., Ogaji, S.O.T., and Probert, S.D. 2006. Reducing the cost of preventive maintenance (PM) through adopting a proactive reliability-focused culture. *Applied Energy*, 83, 1235–1248.

Eti, M.C., Ogaji, S.O.T., and Probert, S.D. 2004. Implementing total productive maintenance in Nigerian manufacturing industries. *Applied Energy*, 79, 385–401.

Gosavi, A. 2006. A risk-sensitive approach to total productive maintenance. *Automatica*, 42, 321–330.

Lévy, J.P. and Varela, M. 2003. *Análisis multivariable para las ciencias socials*. Madrid: Prentice Hall.

Likert, R. 1932. A technique for the measurement of attitudes. *Archives of Psychology*, 22, 1–55.

McKone, K.E., Schroeder, R.G., and Cua, K.O. 1999. Total productive maintenance: a contextual view. *Journal of Operations Management* 17, 123–144.

McKone, K.E., Schroeder, R.G., and Cua, K.O. 2001. The impact of total productive maintenance practices on manufacturing performance. *Journal of Operations Management*, 19, 39–58.

Nunnally, J.C. 1978. *Psychometric Theory*. New York: McGraw-Hill.

Nunnally, J.C. and Bernstein, H. 2005. *Teoría psicométrica*. México: McGraw-Hill Interamericana.

Pinjala, S.K., Pintelon, L., and Vereecke, A. 2006. An empirical investigation on the relationship between business and maintenance strategies. *International Journal of Production Economics*, 104, 214–229.

Pollandt, S. and Wille, R. 2005. Functorial scaling of ordinal data. *Discrete Applied Mathematics*, 147, 101–111.

Reed, R. 1996. Beyond process: TQM content and firm performance. *Academy of Management Review*, 21, 173–202.

Sachdeva, A., Kumar, D., and Kumar, P. 2008. Planning and optimizing the maintenance of production systems in a paper plant. *Computers and Industrial Engineering*, 55, 817–829.

Streiner, D. and Norman, G.R. 1995. *Health Measurement Scales: Practical Guide to Their Development and Use*, 2nd ed. Oxford: Oxford University Press.

Tajiri, M. and Gotoh, F. 1992. *TPM Implementation: A Japanese Approach*. New York: McGraw-Hill.

Takahashi, Y. and Osada, T. 1989. *TPM: Total Productive Maintenance*. Boston: Productivity Press.

Tastle, W.J. and Wierman, M.J. 2007. Using consensus to measure weighted targeted agreement. *Fuzzy Information Processing Society*, 24, 31–35.

Vinodh, S. 2010. Improvement of agility and sustainability: A case study in an Indian rotary switch manufacturing organization. *Journal of Cleaner Production* 18, 1015–1020.

Wikoff, D. 2007. Improve all the M's in TPM system. *Plant Engineering* 61(12), 21–22.

Zhou, W.H. and Zhu, G.L. 2008. Economic design of integrated model of control chart and maintenance management. *Mathematical and Computer Modeling* 47, 1389–1395.

chapter five

Design simplification and innovation through adoption of a time-based design for maintenance methodology

Anoop Desai and Anil Mital

Contents

5.1 Introduction

We live in an era of rapid technological change characterized by globalization. Emerging frontiers and new marketplaces are symbolic of this process. Innovation is the key to survival and prosperity in this dynamic new age. Webster's dictionary defines innovation as novelty or the introduction of something new. It can be appreciated that throughout the history of capitalism, innovation has been the driving force behind economic development and sustenance. Highly innovative and dynamic economies are usually seen as the most prosperous. A cursory look at the past 300 years clearly reveals this fact.

Innovative economies spawn revolutionary new ideas that find expression in new inventions that serve to enhance productivity.

Modern society is always in a state of constant technological evolution. New technologies result in new products that serve to enhance economic productivity. For instance, the nineteenth century was dominated by inventions that improved productivity through the use of mechanical power. The twentieth century achieved the same objective through nuclear power and computing prowess. It is too soon in the twenty-first century to predict with any degree of certainty what new technologies may be developed. What can be said with certainty is that twenty-first century advances will be revolutionary in terms of end results and changes will result from inventions we can only imagine.

New products are direct results of innovation. They are spawned by ideas and inventions. However, innovation alone is not sufficient. Successful products attract competitors. Product design must be dynamic enough to stay one step ahead of the competition and constitutes about 70% of product lifecycle cost.

For that reason, efficient and effective design is an indispensable component of a product lifecycle. Numerous strategies have been devised to enhance various stages of the product design process. One strategy that has been particularly effective is often called Design for X (DfX).

DfX is a design philosophy that seeks to design a product for a specific parameter. Many design objectives applied together constitute effective product design. The X in the term stands for any of the myriad design parameters that must be optimized in the quest for ideal design, for example, designing for ease of assembly, maintenance, cost, or usability. Each of these functions constitutes an individual X component that represents the objective that the methodology seeks to optimize.

When a product is designed for assembly, it possesses certain design characteristics that lend themselves exclusively to the ease with which its component parts can be assembled. Good design means that the time and cost required for assembly are greatly reduced (Li and Hwang 1992, Kim et al. 1995, Boothroyd 1980). The result, however, would be negative if disassembly was the X in question. When a product is designed for disassembly, its design characteristics are very different (Boothroyd 1980 and 1982, Kim et al. 1995). The difference is that the design ensures that disassembly will be facilitated. Products are designed to be disassembled for a variety of reasons, for example, to enable environmentally friendly disposal at the ends of their useful lives.

To succeed in the marketplace, it is also equally important that a product is easy to manufacture at low cost, maintain, repair, and dispose of. Product disposal has gained considerable attention in recent years because of burgeoning demands for consumer products worldwide. Finite amounts of natural resources are unable to keep pace with ever-increasing demands. Resource conservation has therefore become vital.

Maintenance may be seen as a combination of assembly and disassembly processes. Maintenance of a piece of equipment often requires partial disassembly to facilitate the required procedure after which the equipment is re-assembled. Maintenance requirements may cause product designs to change significantly.

Maintenance practices can be classified into two broad categories: corrective and predictive. Every time a device or system fails, a repair or restore action is required to restore it to operability. Equipment failures require corrective maintenance that will restore the system to operation. Corrective maintenance is reactive because it is performed only after a breakdown occurs (Sawhney et al. 2009).

Preventive maintenance, as the term suggests, is carried out to minimize malfunctions and failures. It anticipates future failures and attempts to prevent them by implementing corrective action before they occur (Yeh et al. 2009, Verma and Ramesh 2007). Examples of preventive maintenance tasks are equipment lubrication, part replacements, cleaning and adjustments (e.g., tightening, loosening). Equipment may also be checked for signs of deterioration during preventive maintenance. One hallmark of predictive maintenance is that strict maintenance schedules must be followed if a program is to be truly effective. One of the main advantages of predictive maintenance over preventive maintenance is that equipment is taken offline only when the need is imminent (not on a specific schedule as is the case with preventive maintenance).

Maintenance activities entail significant equipment downtime. Machines that lend themselves to ease of maintenance are important from two perspectives: production and safety. Also, machines that are difficult to maintain routinely are less likely to receive the required standard of maintenance. Thus, it is clear that incorporating maintenance based on the X of the DfX paradigm can result in significant resource optimization. In this chapter, a time-based design for maintenance strategy is described briefly. Design modifications that can occur as a result of the strategy are explained and the effectiveness of this approach is corroborated by a case study.

5.2   Time-based design for maintenance methodology: A brief encapsulation

The principal concepts of the methodology are:

1. The most common maintenance operations are listed and described in complete detail.
2. Each maintenance task is then subdivided into elemental tasks for purposes of simplification.

Note that only a relatively small fraction of all the tasks performed during a maintenance operation are actually responsible for effective maintenance. Examples are reaching for and grasping tools and cleaning components prior to maintenance. Most maintenance activities may be performed onsite or require removal of a component that may constitute a potential cause for failure. If a component must be removed, one or more fasteners may also need removal with ordinary or special tools. Related factors such as the need to exert normal or abnormal force may be explained along similar lines.

Design variables play a crucial role since they affect the performance of the aforementioned tasks. For instance, non-standard fasteners may require specialized tools to affect loosening, unfastening, and refastening (for restoration). Physical features of a component such as weight, shape, and size may also indicate that extra manpower may be needed to perform maintenance. This is especially important in light of the fact that maintenance is a manual operation.

Thus, operations that require extra manpower, involve assumption of unnatural postures for prolonged periods, and entail repetitive motions are obviously not maintenance friendly because additional labor is needed when these operations are performed. Similarly, lifting and carrying requirements (shape, size, weight, and composition) should be considered in component design. Based on these principles, a design for maintenance methodology was formulated.

Most maintenance operations constitute a combination of disassembly, maintenance tasks, and reassembly performed in a specific sequence. For example, a basic disassembly task may require the removal of an easily grasped object—it requires little hand force by a trained worker under average conditions. Such a task corresponds to a score of 73 time measurement units (TMUs). The corresponding time duration of such a task is 2 sec.

The Methods Time Measurement (MTM) system was used to make these calculations. MTM was chosen because it has proven effective and has been adopted widely for a variety of industrial applications. MTM data also eliminates stopwatch studies and other time measurement methods that are inherently subject to high degrees of error based on worker skill levels, accuracy of measuring instruments, and observer experience.

Every maintenance-related task, whether disassembly, task performance, or re-assembly is thoroughly described and analyzed using a numeric index. The total numeric scores pertaining to all tasks are then ranked in descending order. Thus, if a certain task requires 5 sec and another takes 20 sec, the 20-sec task assumes precedence in the task analysis hierarchy.

Figure 5.1 depicts the methodology. Table 5.1 lists the TMUs allocated to each design feature based on the force required to assemble a component. Since all maintenance activities generally entail some disassembly and assembly operations, the methodology considers disassembly as well. Table 5.2 depicts scores assigned to selected disassembly operations.

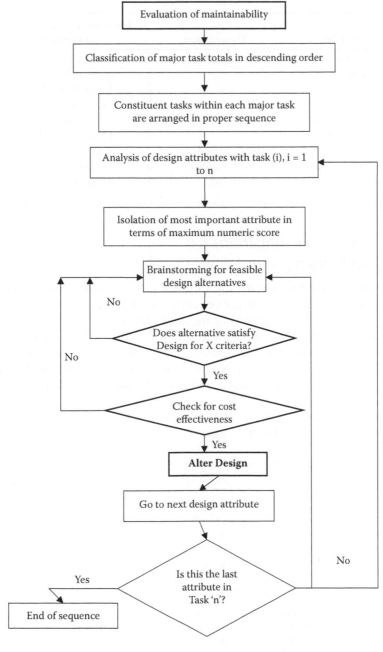

Figure 5.1 Graphic encapsulation of design for maintenance methodology.

Table 5.1 Numerical Scores Assigned to Assembly Operations for Range of Variables

Motion	Description		Force Required to Effect Assembly
Rectilinear motion without pressure	Hand pushing	0.5	Little effort
		1	Moderate effort
		2	Large amount of effort
Rectilinear and curvilinear motions without pressure	Hand twisting and pushing	1	Little effort
		2	Moderate effort
		4	Large amount of effort
Rectilinear and curvilinear motions with exertion of pressure	Intersurface friction and/or wedging	2.5	Little effort
		3	Moderate effort
		5	Large amount of effort

5.3 Design modifications for simplifying maintenance

After design attributes with high numeric scores have been identified for each component as described above, causal effects must be diagnosed. An in-depth diagnosis of these effects may lead to the development of alternative design configurations. Thus, the design diagnostics part of the algorithm is indispensable for effective application of the methodology. A few design diagnostics for various design attributes (accessibility, tool positioning, conditions of mating surfaces, etc.) are cited in Table 5.3.

It is clear from Table 5.3 that numerous design simplifications can promote ease of maintenance assembly and disassembly. This is made

Table 5.2 Numerical Scores Assigned to Disassembly Operations for Range of Variables

Design Attribute	Design Feature	Design Parameters	Score	Significance
Force required to affect disassembly	Rectilinear motion without exertion of pressure	Push/pull operations with hand	0.5	Little effort
			1	Moderate effort
			3	Large amount of effort
	Rectilinear motion with exertion of pressure	Intersurface friction and/or wedging	2.5	Little effort
			3	Moderate effort
			5	Large amount of effort

possible by following the methodology described in Figure 5.1 in conjunction with the statistics presented in Table 5.1. Design simplifications are
defined as features that enable the easier performance of a certain activity such as assembly, disassembly, or maintenance in less time. Thus, it is
logical that a certain fastening device such as a snap fit would require less
activity than manipulating a screw-in joint.

The choice of an alternating fastening technique depends on the outcome of a feasibility study that evaluates the mechanical, structural, and
cost aspects of a proposed change. Table 5.4 shows the evaluation of a
screw-in joint. Table 5.5 provides a comparison by evaluating a snap-fit
joint. Clearly, based on the sizes, dimensions, and locations of typical
industrial screws, removing or fitting a screw in place is often time consuming and may be very tiring.

Thus, based on the methodology presented in this chapter, it will
be observed that typical screw-in joints do not promote operational efficiency. Thus, they can be and often are replaced with types of joints that
incorporate other fastening techniques based on the results of thorough
engineering and cost analyses.

We can see that a typical snap-fit joint is substantially easier to disas-
semble than most other types of joints. The degree of ease is reflected in
the amounts of force and time required to perform a maintenance operation as shown in a comparison of Tables 5.4 and 5.5. Disassembly of a
snap-fit joint in a typical consumer product requires about 8.70 TMUs
compared to 15.15 TMUs for a screw-in joint. This represents a time saving of almost 58% for a snap-fit joint.

The following section presents a case study of a real product that
illustrates the time savings that can be accomplished by using the design
diagnostics presented in this chapter.

5.4   Case study: Lubrication of a hand-held drill

The maintenance operation described in this section is lubrication of the
rotor of a typical hand-held drill with screw-in joints. Table 5.6 demonstrates the sequence of operations necessary to open the drill housing,
expose the rotor, lubricate it, and re-assemble the device. A total of eight
screw-in joints make lubrication a very tedious operation.

A postmaintenance design diagnostic revealed that removing all the
screws was the most time-consuming part of the lubrication task. An
engineering analysis revealed that the two middle screw-in joints in the
original product were functionally essential. However, the remaining six
screws could be replaced with snap fits without jeopardizing the structural integrity and functionality of the product.

Table 5.7 demonstrates the maintenance process of the product after
the design modification. Clearly, the design modifications made after the

Table 5.3 Design Diagnostic Tool for Achieving Design Simplification for X

Design Attribute	Design Feature	Remedial Measures	Component Redesign Required?
Accessibility	Deep fastener recesses	Redesign recesses to facilitate tool access	Y
		Select different fastening method	Y
	Narrow fastener recesses	Redesign recesses to facilitate tool access	Y
		Select different fastening method	Y
	Small fastener head	Increase fastener head size	N
		Select different fastening method	Y
	Obscure fastener	Choose standard fastener sizes	N
		Increase fastener size	N
		Select different fastener method	Y
	Deformed component	Improve component rigidity to withstand stress	Y
		Redesign weak component cross sections	Y
	Need for cleaning before access	Redesign component–fastener interface	Y
		Change component material	Y
	Obscuring components	Redesign assembly and/or disassembly sequence	N
Force Exertion	Moderate to large force required	Select appropriate materials for component bearing surfaces and/or fastener to reduce friction	Y
		Redesign component holding surfaces	Y
	Tight snap fits	Redesign components to provide adequate clearance and taper to allow easy dismantling of snap fits	Y
	Coarse threads on fasteners	Select fasteners with finer threads or greater thread pitch	N

(continued)

Table 5.3 Design Diagnostic Tool for Achieving Design Simplification for X
(continued)

Design Attribute	Design Feature	Remedial Measures	Component Redesign Required?
Positioning	Moderate to high degree of precision required to place tool	Redesign access path Modify component bearing surfaces and/or fasteners	Y
	Component size	Use standard sizes	Y
		Optimize component size functionality and material handling	Y
	Component shape	Use standard and/or symmetric shapes	Y
		Eliminate component protrusions	Y
Mating surface condition	Either of all surfaces corroded	Select appropriate non-corrosive materials for component bearing surfaces and/or fasteners	Y
	Either or all surfaces deformed	Select appropriately rigid materials to withstand forces during assembly and disassembly	Y
		Redesign component bearing surfaces to allow appropriate clearances	Y
		Redesign fastener holding surfaces	N

Y = positive response. N = negative response.

diagnostic test resulted in a far better product from a maintenance perspective. The reader should bear in mind that the main design modification was replacement of six screw-in joints with six snap fits without violating the structural integrity and functionality of the product.

By comparing Tables 5.6 and 5.7, we can see that total maintenance time decreased from 1.764 min to 1.21 min and the number of operations was reduced as well. A time saving of 31% was achieved while the structural and functional characteristics of the product were retained. In addition, the product became easier to handle due to weight reduction and easier to recycle due to reductions in material variability.

Table 5.4 Numeric Analysis of Typical Screw-In Joint during Assembly (TMUs)

Task Total	Intersurface Friction	Component Size	Component Weight	Component Symmetry	Torque Exertion	Dimensions	Location	Accuracy of Tool Placement	Visual Fatigue Allowance
15.12	2	2	2	0.8	2	1.6	2	2	5%

Table 5.5 Numeric Analysis of Typical Snap-Fit Joint during Disassembly (TMUs)

Task Total	Intersurface Wedging	Component Size	Component Weight	Component Symmetry	Force Exertion	Dimensions	Location	Accuracy of Tool Placement	Visual Fatigue Allowance
8.70	1	3.5	2	1.4	1	1	1.6	1.6	1%

5.5   Conclusion

This chapter presented a comprehensive methodology to positively affect product maintenance. Time was used as the singular metric of measurement. Metrics for measuring other parameters such as ease of maintenance are available but are beyond the scope of this chapter. A design diagnostic to evaluate various design parameters was also presented. Such diagnostic measures are essential because they reveal means to alter current designs to enhance product features. Finally, a case study was presented to demonstrate the efficacy of the diagnostic methodology based on a before-and-after design change scenario for a common consumer product.

Table 5.6 Maintenance Operation: Predesign Modification for Lubrication of Drill Rotor

Task No.	Task Description: Lubrication of Drill Rotor in Unmodified Product	Task Total	(Dis)Assembly Force			Material Handling			Tooling		Accessibility & Positioning			Allowances			
			Intersurface Friction	Intersurface Wedging	Material Stiffness	Component Size	Component Weight	Component Symmetry	Force Exertion	Torque Exertion	Dimensions	Location	Accuracy of Tool Placement	Posture Allowance	Motions Allowance	Manpower Allowance	Visual Fatigue Allowance
1	**Remove upper housing**																
1a	Unscrew 1st of 6 front/back screws	15.65	2.5	–	–	2	2	0.8	–	2	1.6	2	2	–	–	–	5%
1b	Unscrew 2nd of 6 front/back screws	15.65	2.5	–	–	2	2	0.8	–	2	1.6	2	2	–	–	–	5%
1c	Unscrew 3rd of 6 front/back screws	15.65	2.5	–	–	2	2	0.8	–	2	1.6	2	2	–	–	–	5%

(continued)

Table 5.6 Maintenance Operation: Predesign Modification for Lubrication of Drill Rotor (continued)

Task No.	Task Description: Lubrication of Drill Rotor In Unmodified Product	Task Total	(Dis)Assembly Force			Material Handling			Tooling		Accessibility & Positioning			Allowances			
			Intersurface Friction	Intersurface Wedging	Material Stiffness	Component Size	Component Weight	Component Symmetry	Force Exertion	Torque Exertion	Dimensions	Location	Accuracy of Tool Placement	Posture Allowance	Motions Allowance	Manpower Allowance	Visual Fatigue Allowance
1d	Unscrew 4th of 6 front/back screws	15.65	2.5	–	–	2	2	0.8	–	2	1.6	2	2	–	–	–	5%
1e	Unscrew 5th of 6 front/back screws	15.65	2.5	–	–	2	2	0.8	–	2	1.6	2	2	–	–	–	5%
1f	Unscrew 6th of 6 front/back screws	15.65	2.5	–	–	2	2	0.8	–	2	1.6	2	2	–	–	–	5%
1g	Unscrew 1st of 2 middle screws	15.65	2.5	–	–	2	2	0.8	–	2	1.6	2	2	–	–	–	5%

1h	Unscrew 2nd of 2 middle screws	2.5	15.65	—	—	2	2	—	0.8	2	1.6	2	2	—	—	5%
1i	Pull out upper housing	—	8.7	1	—	3.5	2	1	1.4	—	1	1.6	1.6	—	—	1%
2	**Access drill rotor**															
2a	Pull out bushing	1	10.5	—	—	2	2	1	0.8	—	1	1	1.2	—	—	5%
2b	Pull out insulating washer	1	10.5	—	—	2	2	1	0.8	—	1	1	1.2	—	—	5%
3	**Lubricate rotor and clean housing**															
3a	Lubrication: 2/ location × 2 locations	—	4.2	—	—	—	—	—	—	—	—	—	—	—	—	5%
3b	Clean upper housing; 2	—	2.02	—	—	—	—	—	—	—	—	—	—	—	—	1%
4	**Re-assemble drill**															
4a	Re-install washer	1	10.5	—	—	2	2	1	0.8	—	1	1	1.2	—	—	5%
4b	Re-install bushing	1	10.5	—	—	2	2	1	0.8	—	1	1	1.2	—	—	5%

(continued)

Table 5.6 Maintenance Operation: Predesign Modification for Lubrication of Drill Rotor (continued)

Task No.	Task Description: Lubrication of Drill Rotor in Unmodified Product	Task Total	(Dis)Assembly Force			Material Handling			Tooling		Accessibility & Positioning			Allowances			
			Intersurface Friction	Intersurface Wedging	Material Stiffness	Component Size	Component Weight	Component Symmetry	Force Exertion	Torque Exertion	Dimensions	Location	Accuracy of Tool Placement	Posture Allowance	Motions Allowance	Manpower Allowance	Visual Fatigue Allowance
4c	Re-fit upper housing	8.7	–	1	–	3.5	2	1.4	1	–	1	1.6	1.6	–	–	–	1%
4d	Screw 1st middle screw	15.65	2.5	–	–	2	2	0.8	–	2	1.6	2	2	–	–	–	5%
4e	Screw 2nd middle screw	15.65	2.5	–	–	2	2	0.8	–	2	1.6	2	2	–	–	–	5%
4f	Screw 1st front/ back screw	15.65	2.5	–	–	2	2	0.8	–	2	1.6	2	2	–	–	–	5%
4g	Screw 2nd front/back screw	15.65	2.5	–	–	2	2	0.8	–	2	1.6	2	2	–	–	–	5%
4h	Screw 3rd front/back screw	15.65	2.5	–	–	2	2	0.8	–	2	1.6	2	2	–	–	–	5%

4i	Screw 4th front/ back screw	15.65	2.5	–	2	2	0.8	–	2	1.6	2	2	–	–	–	5%
4j	Screw 5th front/ back screw	15.65	2.5	–	2	2	0.8	–	2	1.6	2	2	–	–	–	5%
4k	Screw 6th front/ back screw	15.65	2.5	–	2	2	0.8	–	2	1.6	2	2	–	–	–	5%
		294	Total time for maintenance operation: 2940 TMUs = 1.764 min												

Table 5.7 Maintenance Operation: Postdesign Modification for Lubrication of Drill Rotor

Task No.	Task Description: Lubrication of drill rotor in modified product	Task Total	(Dis) Assembly Force			Material Handling			Tooling		Accessibility & Positioning			Allowances			
			Intersurface Friction	Intersurface Wedging	Material Stiffness	Component Size	Component Weight	Component Symmetry	Force Exertion	Torque Exertion	Dimensions	Location	Accuracy of Tool Placement	Posture Allowance	Motions Allowance	Manpower Allowance	Visual Fatigue Allowance
1	**Remove upper housing**																
1a	Snap open first snap-fit joint	10.91	–	1	–	3.5	1	1	1	–	1	1.2	1.6	–	–	–	1%
1b	Snap open second snap-fit joint	10.91	–	1	–	3.5	1	1	1	–	1	1.2	1.6	–	–	–	1%
1c	Snap open third snap-fit joint	10.91	–	1	–	3.5	1	1	1	–	1	1.2	1.6	–	–	–	1%
1d	Snap open fourth snap-fit joint	10.91	–	1	–	3.5	1	1	1	–	1	1.2	1.6	–	–	–	1%
1e	Snap open fifth snap-fit joint	10.91	–	1	–	3.5	1	1	1	–	1	1.2	1.6	–	–	–	1%

1f	Snap open sixth snap-fit joint	10.91	–	1	–	3.5	1	1	1	1	–	1	1	1.2	1.6	–	1%
1g	Unscrew 1st of 2 middle screws	15.65	2.5	–	2	2	2	0.8	–	2	1.6	2	2	–	–	–	5%
1h	Unscrew 2nd of 2 middle screws	15.65	2.5	–	2	2	2	0.8	–	2	1.6	2	2	–	–	–	5%
1i	Pull out upper housing	8.7	–	1	–	3.5	2	1.4	1	–	1	1.6	1.6	–	–	–	1%
2	**Access drill rotor**																
2a	Pull out bushing	10.5	1	–	2	2	2	0.8	1	–	1	1	1.2	1.2	–	–	5%
2b	Pull out insulating washer	10.5	1	–	2	2	2	0.8	1	–	1	1	1.2	1.2	–	–	5%
3	**Lubricate rotor and clean housing**																
3a	Lubrication: 2/location × 2 locations	4.2	–	–	–	–	–	–	–	–	–	–	–	–	–	–	5%
3b	Clean upper housing: 2	2.02	–	–	–	–	–	–	–	–	–	–	–	–	–	–	1%
4	**Re-assemble drill**																
4a	Re-install washer	10.5	1	–	2	2	2	0.8	1	–	1	1	1.2	1.2	–	–	5%
4b	Re-install bushing	10.5	1	–	2	2	2	0.8	1	–	1	1	1.2	1.2	–	–	5%
4c	Re-fit upper housing	8.7	–	1	–	3.5	2	1.4	1	–	1.6	1.6	1.6	1.6	–	–	1%
4d	Screw 1st middle screw	15.65	2.5	–	2	2	2	0.8	–	2	1.6	2	2	2	–	–	5%
4e	Screw 2nd middle screw	15.65	2.5	–	2	2	2	0.8	–	2	1.6	2	2	2	–	–	5%
4f	Snap all 6 joints in place	7.58	–	1	–	3.5	1	1	1	–	–	–	–	–	–	–	1%
		201.3		Total time for maintenance operation: 2013 TMUs = 1.21 min													

References

Boothroyd, G. 1982. Design for assembly: the road to higher productivity. *Assembly Engineering*, March.

Boothroyd, G. 1980. *Design for Assembly: A Designer's Handbook.* Amherst: University of Massachusetts Press.

Kim, G., Lee, S., and Bekey, G. 1995. Comparative assembly planning during assembly. *Computers in Industrial Engineering*, 22, 403–413.

Li, R. and Hwang, C. 1992. "A framework for automatic DFA system development," *Computers in Industrial Engineering*, 22 (4), 403–413.

Sawhney, R., Kannan, S., and Li, X. 2009. Developing a value stream map to evaluate breakdown maintenance operations. *International Journal of Industrial and Systems Engineering*, 4, 229–240.

Verma, A.K. and Ramesh, P.G. 2007. Multi-objective initial preventive maintenance scheduling for large engineering plants. *International Journal of Reliability, Quality, and Safety Engineering*, 14, 241–250.

Yeh, R.H., Kao, K.C., and Chang, W.L. 2009. Optimal preventive maintenance policy for leased equipment using failure rate reduction. *Computers and Industrial Engineering*, 57, 304–309.

chapter six

Changes in manufacturing worker productivity and working hours in the United States

Vignesh Ravindran, Aashi Mital, and Anil Mital

Contents

6.1   Introduction

Productivity is widely defined as the ratio of output produced to the inputs or resources required to produce it (Giampietro et al. 1993). The output produced may vary based on industry, for example, household goods, clothing, electrical devices, and even medical services offered by doctors to patients. The inputs required to produce outputs may consume time, labor, capital, materials, and machine use. Productivity can be interpreted in many ways across various disciplines; they all generate valuable socioeconomic insights. A productivity study has no specific goals per se. Depending on the subject and need to understand a specific concept, the input and output parameters may be selected accordingly.

The objective of this chapter is to discuss U.S. workforce productivity in manufacturing industries and reveal how it has changed over the years as workweeks have shortened.

Traditionally, the *labor* term describes unskilled manual work. The modern definition of productivity, however, has evolved to include the contributions of all skilled and unskilled workers who earn wages. Labor productivity is the amount of goods and services that a worker produces in a given unit of time and indicates resourcefulness in an industry.

Productivity may be improved in two ways: (1) by increasing the output produced, or (2) by decreasing the input required to produce the specified output. In any productivity study, the choice of input and output parameters is vital to meeting its research objectives. One of the most significant aspects of a productivity study is quantification of labor input. It is difficult to quantify this input because each worker is inherently different from others. The differences may be based on skill, the nature of work, desire, dedication, experience, training, or other issues. No specific way exists to estimate the direct contribution of a worker to productivity. One can safely assume, however, that all workers, regardless of their skills and ability, make significant contributions to a company during their working hours.

The use of working hours as an input parameter conveniently collects all worker contributions under the notion of time. In this chapter, labor productivity (output) of workers in manufacturing industry per unit working hour (input) is discussed. Productivity trends highlight a country's real progress, identify technological trends, depict standards of living, and predict future economic growth. The value of analyzing labor productivity trending as an output parameter cannot be overemphasized.

The discussion is divided into three parts: (1) shift in labor force away from agriculture to industry, (2) changes in working hours over the past 100 years, and (3) labor productivity trends since 1900.

6.2 Shifts in labor workforce

To understand the changes in industrial labor productivity within the United States., we must first recognize the nature of the workforce that contributed to industrial productivity and its transformations over several decades. The U.S. workforce grew from nearly 24 million in 1900 to approximately 140 million in 2000.

During that time, a massive transition from an agriculture-based society to an industrial society occurred and the workforce shifted accordingly. This change created a foundation based on economic growth and development that ultimately shaped modern U.S. society. The shift in labor workforce, its relationship with productivity, and effects on working hours are discussed in the following subsections.

6.2.1   Onset of the industrial revolution

The Industrial Revolution is considered the main reason behind the deterioration of an agricultural economy. Until then, agriculture was the staple activity of human civilization worldwide. The food produced allowed humans to endure and survive and also paved the way to stability and growth. Most production activities took place within the domestic sphere. Many workers spent a great deal of time, skill, and effort to produce goods that were incredibly rare or expensive such as linens and clothing. In the eighteenth century, however, Europe underwent a production and market revolution that started in Great Britain, permeated continental Europe and eventually reached North America's shores.

The United States was influenced greatly by the European industrialization. Samuel Slater, known as the father of the Industrial Revolution in America, introduced manufacturing technology in his cotton mills in Massachusetts during the 1790s. Eli Whitney's cotton gin, Elias Howe's sewing machine, and Robert Fulton's steamboat are considered the triggers of American industrialization.

Despite its tremendous popularity and success in Europe, the industrial movement lacked the same momentum in North America because of reliance upon farming. As citizens of a new world, North Americans were more interested in exploring their abundant resources rather than investing capital in machine-based manufacturing. The Embargo Act of 1807 and the War of 1812 were the first events that sparked the U.S. Industrial Revolution (Kelly 2011).

The Embargo Act closed all U.S. exports to Europe, effectively throttling overseas trade. The War of 1812 clearly demonstrated the need for the United States to increase its industrial productivity if it wanted to achieve economic independence. The need to modernize and industrialize what was predominantly an agricultural workforce was recognized. As a result, capital investments were made in local manufacturing and textile industries. Transportation and communication were improved and widespread use of electricity produced further momentum.

U.S. industrialization reached a turning point during the Civil War in the 1860s. The war was literally a technological race between the fighting states. The end of the war brought tremendous growth of manufacturing and technological advances. After World War II in the middle of the twentieth century, the United States emerged as one of the most powerful industrial nations in the world.

Three important changes signified the advent of the Industrial Revolution: (1) machines based on the concept of automating tools, (2) steam power as a replacement for hard human labor, and (3) mass production through dedicated plants and assembly lines. Workers could make more products with the help of machinery powered by external sources like steam and electricity.

Industrialization thus added a new dimension to productivity. As industrialization gained momentum, production volume and variety increased. Productivity (number of items a worker produced) in a manufacturing plant with the help of machinery far exceeded the productivity of an average farmer who depended heavily on manual labor. In other words, industrialization led to increased labor productivity. As productivity increased, workers' ability to purchase goods increased. More purchasing capacity then increased consumer needs and purchases. As a result, the Industrial Revolution started and also promoted an endless cycle of need and consumption. To date, the cycle continues to drive U.S. society toward higher standards of living.

U.S. agriculture also underwent a tremendous transformation from a labor-intensive pursuit in the early twentieth century to a technology-intensive industry by the twenty-first century due to mechanization and industrialization.

6.2.2   Decline of agriculture

The introduction of farm mechanization in the early twentieth century changed the landscape of U.S. agriculture. Small farms growing diversified crops around the country transformed into a small number of large specialized farms concentrated in specific geographical regions. This transformation exerted a major socioeconomic impact on society as mechanized farm operations replaced manual labor with machines. This replacement marked the decline of the agricultural workforce in the United States. At the end of the nineteenth century, nearly 50% of the population was involved in agricultural employment. By the end of the twentieth century, a rapid and steady decline in the agricultural workforce meant that less than 2% of the U.S. population worked in agriculture. See Figure 6.1.

In the past, agriculture was a rural activity and farming families formed communities. As the mechanization of farming gained momentum, the agricultural workforce decreased because machinery led to greater production with fewer workers. The concept of farming in families gradually diminished over time and machinery allowed farmers to increase the sizes of their farms and the crops they produced. The increase

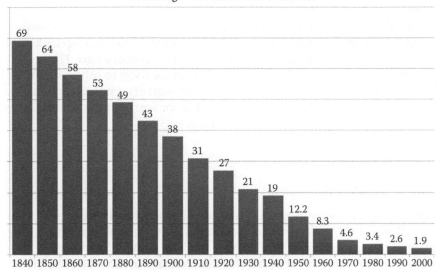

Percentage of Farmers in US Workforce

Figure 6.1 Migration of U.S. workforce from farms from 1840 to 2000. According to the U.S. Census Bureau (1952), 1950 was considered the beginning of agricultural mechanization and use of improved agricultural techniques. (Source: Dimitri, C., Effland, A., and Conklin, N. 2005. *U.S. Department of Agriculture Economic Research Service Economic Information Bulletin 3.*)

in farm size pushed farmers away from dense populations and big cities. The disappearance of farming communities meant that farmers worked in isolation without community and social amenities.

Farming gradually lost its appeal as a way to make a living and families started searching for jobs in cities. Farmers needed to make massive capital investments in expensive machines and technology to remain competitive. Agricultural investments grew increasingly less attractive. These changes caused the migration of workers away from agriculture.

The impact of technology on U.S. farming cannot be minimized (Holt 1970). Machine-driven farms, development of effective pesticides and fertilizers, genetically modified crops, and advances in animal breeding are examples that highlight the impact of technology on farming. The high levels of productivity of these technology-intensive farms made U.S. agriculture increasingly efficient and stable despite the decrease in the size of its workforce.

Over time, the average American family has spent less money on food and more on other purchases and activities, thus aiding national development. The decline of the agricultural workforce also meant a shift in employment to other occupations. This shift away from agriculture meant that the workforce no longer concentrated on producing food and could pursue other activities that fostered economic development in the United States.

Farming was seen as a vital factor in human survival; it is now perceived as another contributor to a major economy. Modern agriculture's future impacts are far smaller than they were a century ago. The contribution of farming to modern economy in comparison to other industries has decreased steadily as has the number of workers employed on farms. Understanding this shift in the nature of labor workforce is necessary to comprehend changes in the labor workweek.

6.3   Working hours

6.3.1   Nineteenth century

The United States was an agriculture-based economy in the nineteenth century. Before 1900, most workers performed manual work on farms. They planted seeds, walked behind plows, milked animals, and forked hay; some farmers were helped by horses (U.S. Census Bureau 1950). The manufacturing industries of the 1830s depended on a combination of farming, textile production, transportation, and electricity (Kelly 2011). Although the incipient technology used in early industry operated with basic farm labor, a lot of man-hours were required to increase production. The productivity of a nineteenth-century worker was assumed to be directly proportional to the number of hours he committed to physical labor.

Figure 6.2 clearly indicates that most production workers in the nineteenth century worked 10 to 11 hours daily and sometimes worked up to 13 hours per day. The average workweek was 55 to 60 hours (Figure 6.3; Atack and Bateman 1992).

6.3.2   The shorter-hours movement from the late eighteenth to early twentieth centuries

The battle for shorter working hours started in May 1791, when Philadelphia carpenters carried out the first unsuccessful labor strikes for 10-hour workdays (Whaples 2011, Hunnicutt 1984). In 1845, after the agitation by mill workers in Lowell, Massachusetts, the 10-hour workday concept gained popularity. In 1874, Massachusetts enforced the first 10-hour law for female workers.

By 1880, the average daily workday across all industries stabilized at 10 hours as seen in Figure 6.2. The labor unions pushed further for 8-hour workdays. Several strike attempts in support of 8-hour workdays in the 1880s failed tragically (e.g., the Haymarket Square bombing), mainly due to weak conviction and the poor organization skills of a union known as the Knights of Labor.

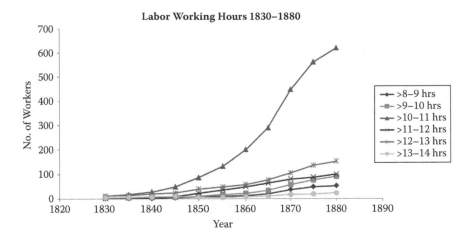

Figure 6.2 Number of workers and the average daily hours worked in manufacturing from 1830 to 1880. The line with triangle dots indicates the number of workers working more than 10 to 11 hours per day. (Source: Carter, S.B. et al. 2011. *Historical Statistics of United States Millennial Edition Online.* Cambridge University Press. With permission.)

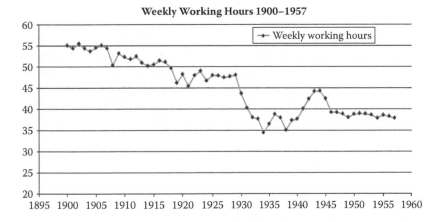

Figure 6.3 Average weekly hours worked in manufacturing from 1900 to 1957. The workweek decreased from 55 hours in the 1900s to about 40 per week In the 1950s. The graph shows a sharp dip in working hours during the Great Depression. (*Source:* Atack, B. and Bateman, F. 1992. *Journal of Economic History,* 21(1), 129–160. With permission.)

By 1886, the American Federation of Labor (AFL) had been formed and one of its priorities was a shorter workday. The early twentieth century saw a steady decline in working hours. The high demand for workers during World War I provided labor unions with much-needed bargaining power to push for reduced working hours. The 1912 Federal Public Works Act, Senator LaFollette's 1913 bill, and the Adamson Act of 1916 all focused on 8-hour workdays across all industries.

The 55-hour working week in 1900 was gradually reduced to a 50-hour week by the 1920s. The fight for shorter hours reached its peak in the 1930s when the AFL pushed for 6-hour workdays and 5-day workweeks through the Black and Connery bills. For a brief period in the 1930s, the combined effect of economic depression and persistent union pressure for shorter hours led to the 36-hour workweek—the shortest in U.S. history.

6.3.3   Forty-hour workweek and Fair Labor Standards Act

In the 1920s, the U.S. economy was plagued by overproduction and declining consumption. The AFL voiced demands for shorter hours as a cure for growing unemployment and overproduction on the basis that shorter hours would decrease physical demands, share work, spread employment, reduce excess production, and ensure fair standards of living for everyone. Businessmen condemned leisure as an ethical disease and believed that shorter hours wasted human capacity and stemmed economic progress. Their solution to overproduction was to stimulate demand and thus increase consumption (Hunnicutt 1984).

"Consumption economists" identified the bridge between the shorter-hours demands of labor and the increased-consumption demands of business. They believed that reduced working hours implied increased leisure; increased leisure led to increased consumption and was thus good for business. Henry Ford took the first step in that direction in 1914 by decreasing working hours and increasing wages. The concept of promoting consumption through reduced working hours gradually gained industrial appeal.

However, excessive leisure reduced economic capacity. Reduced economic capacity was not appealing because it kept workers from increasing consumption in their leisure time. This imbalance of leisure and desire for consumption was the primary reason behind the demise of the 6-hour workday. There was a growing belief that free time was an economic tragedy that had to be eliminated by increasing economic activity.

The Roosevelt administration in 1933 used shorter hours as a ruse to allay unemployment fears and later dismissed shorter hours as a symbol of economic depression forced upon workers. The administration's plan for economic recovery from the Depression was known as the New Deal.

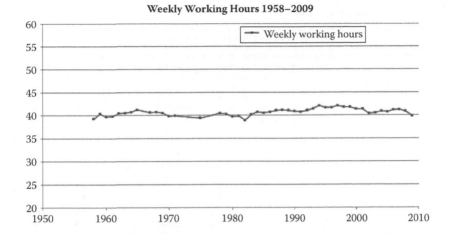

Figure 6.4 Average weekly hours worked in manufacturing from 1958 to 2009. The industrial standard of the 8-hour workday and 5-day workweek has existed for the past half century. (Source: U.S. Census Bureau, 2011.)

It was based on government spending to stimulate demand and guaranteed workweeks. Its objectives of recovering from the Depression and ending unemployment were hugely appealing to the public. The New Deal culminated in the popular Fair Labor Standards Act (FLSA) in 1938 that legally defined a workweek as 44 hours, and later reduced it to 40 (Jones 1975, Rones et al. 1997). The workweek in U.S. industries after the 1940s averaged approximately 40 hours in compliance with the FLSA.

6.3.4   Working hours in the late twentieth century

In contrast to the labor struggles early in the twentieth century, the second half of the century was rather quiet (Figure 6.4). The past 60 years witnessed a tremendous growth in new industries and fast-paced changes in traditional industries although the workweek remained basically unchanged around the 40-hour mark (Figure 6.5). The fight for shorter workweeks reached a 40-hour plateau in the 1950s and has remained there.

Americans have generally embraced the 40-hour workweek that allows them to maintain a reasonable balance between personal life and work. Sharp increases in productivity (discussed in the following sections) have not impacted working hours. The increasing standard of living in the United States and increasing desire for consumer goods helped maintain the 40-hour workweek despite increased productivity. Workers have accepted the shorter workweek over additional income.

Figure 6.5 (U.S. Census Bureau 2012) shows the distribution of workers across industries based on working hours. It clearly summarizes the

Persons at Work by Hours Worked: 2010

[In thousands (134,004 represents 134,004,000), except as indicated. Annual averages of monthly figures. Persons "at work" are a subgroup of employed persons "at work," excluding those absent from their jobs during reference period for reasons such as vacation, illness, or industrial dispute. Civilian noninstitutionalized population 16 years old and over. Based on Current Population Survey; see text, Section 1, and Appendix III. See headnote Table 606, regarding industries]

Hours of Work	Persons at Work (1,000)			Percent Distribution		
	Total	Agriculture and Related Industries	Non-agricultural Industries	Total	Agriculture and Related Industries	Non-agricultural Industries
Total	**134,004**	**2,113**	**131,891**	**100.0**	**100.0**	**100.0**
1 to 34 hours	35,097	592	34,505	26.2	28.0	26.2
1 to 4 hours	1,559	53	1,506	1.2	2.5	1.1
5 to 14 hours	5,488	137	5,351	4.1	6.5	4.1
15 to 29 hours	17,272	260	17,012	12.9	12.3	12.9
30 to 34 hours	10,778	142	10,636	8.0	6.7	8.1
35 hours and over	98,907	1,521	97,386	73.8	72.0	73.8
35 to 39 hours	9,695	111	9,584	7.2	5.3	7.3
40 hours	56,478	591	55,886	42.1	28.0	42.4
41 hours and over	32,734	818	31,916	24.4	38.7	24.2
41 to 48 hours	11,370	152	11,218	8.5	7.2	8.5
49 to 59 hours	12,530	238	12,292	9.4	11.3	9.3
60 hours and over	8,834	428	8,406	6.6	20.3	6.4
Average weekly hours: Persons at work	38.2	41.8	38.1	(X)	(X)	(X)
Persons usually working full-time[1]	42.2	47.7	42.2	(X)	(X)	(X)

X Not applicable. [1]Full-time workers are those who usually worked 35 hours or more (at all jobs).
Source: U.S. Bureau of Labor Statistics, "Employment and Earnings Online," January 2011 issue, March 2011, <http://www.bls.gov/opub/ee/home.htm> and <http://www.bis.gov/cps/home.htm>.

Figure 6.5 Workers and hours worked. (*Sources:* U.S. Census Bureau, 2012 and U.S. Bureau of Labor Standards Report for 2010.)

shift of the U.S. workforce away from agriculture. Only 1.6% of the total working population is involved in farming and farm-based industries as of 2010. The highlighted box in the figure indicates that 72% of the total population works 35 hours or more; most (66%) work about 41 hours a week. Figure 6.5 also confirms that the distribution is the same across both agriculture and non-agricultural industries.

6.3.5   Working hours in the twenty-first century

The early twenty-first century is witnessing a new shift in labor composition: a steady decline in the percentage of labor working in primary and secondary goods producing occupations. By 2000, the working population employed in primary occupations (farming, fishing, and forestry) was less than 3%. Only 19% of the total working population was employed in secondary occupations (manufacturing, mining, and construction). Nearly 78% of the U.S. working population is employed in the service sector. The twenty-first century workweek is greatly influenced by the service sector and is steadily increasing beyond the 40-hour mark. The effects of a service economy on working hours and productivity are discussed briefly late in this chapter.

6.4   Productivity studies

Productivity, although simple by definition, is an inherently complex concept when it comes to interpretation. Productivity studies covering all industries in the manufacturing sector are extremely complicated due to natures and magnitudes of variations involved and the tremendous volume of data to be gathered, analyzed, interpreted and, most importantly, presented in usable form. The types of products and commodities govern the nature of each industry in the manufacturing sector.

Differences in technologies and techniques also exist among industries producing similar products (Kendrick 1956). Industries using the same techniques may also differ in capital investments in technology and equipment. Differences in capital investments create variations in quality and differences in productivity. Industrial productivity also differs based on management decisions, policies, and strategies. The number of factors influencing productivity is much too large to be discussed in detail. We will therefore focus on the contributions of labor to productivity.

The contribution of labor to productivity is difficult to measure because it is a function of several internal and external factors. Internal factors shape worker activity. External factors are beyond worker control but still influence productivity. Internal factors impacting labor include education, skill, interest in quality, cultural beliefs, behavior, attitude, societal trends, and ethics. External factors affecting labor are the nature of work, technology, geographic location of work, plant environment, flexibility, control, management policies, training, salary and benefits, bonuses and other performance rewards, and motivation. Figure 6.6 shows some of these factors.

6.4.1   Importance of labor productivity

Labor productivity is an important determinant of standard of living. As noted earlier, the more productive a labor force, the more efficient the use of resources and the greater the profits. A management decision to share profits with workers results in a general increase in worker income levels, promotes consumption, and thus improves overall standards of living.

Labor productivity indicates the efficiency with which labor uses resources in production. Lower productivity implies waste of resources. Sustained low productivity is not good for the economy. It is therefore imperative that labor productivity be increased by extracting the most benefit from available resources.

There is a physical limit to a worker's capacity to produce. Supplanting a worker with technology circumvents the worker's inherent limitations. Labor productivity provides a general measure of the benefits from science and technology. Greater productivity arises from the use of more

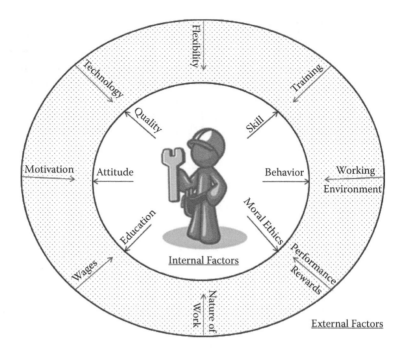

Figure 6.6 Factors affecting labor productivity.

advanced technology to assist labor in production. Labor productivity studies explain the impacts of technology; sectors of low productivity point to the need for investment in technology and developmental research.

Labor productivity trends of an industry also reflect the current management trends. Combining labor, capital, technology, and resources is the responsibility of an industrial management team. Only an efficient combination will generate an increase in productivity. Labor productivity reflects the policies used and decisions made by management in an effort to increase productivity. A decrease in productivity indicates a need to evaluate options for better management methods and policies.

The manufacturing sector remains an important area of the U.S. economy but it now lags behind manufacturing in emerging Asian economies such as China and India. This is mainly due to low labor costs that tip the scales in favor of Asian enterprises. U.S. companies often develop products at the higher end of the technology spectrum and prefer to manufacture them in Asia to save labor costs. This offshoring (or outsourcing) of manufacturing jobs to Asia has reduced employment opportunities in the United States and impacted income distribution.

Furthermore, U.S. makers of consumer products at the lower end of the technology spectrum are finding it increasingly difficult to penetrate markets against their Asian counterparts. The cost of these products is based more on labor cost than on the technology used in production. Hence, the Asian products available to U.S. consumers are less expensive than domestic products without compromising quality.

Most consumer products used routinely fall into the low-tech category and it is important for U.S. industry to recapture the domestic market for such products. Improving labor productivity is a vital factor in invigorating its manufacturing sector over a wide range of products. To recover lost manufacturing jobs and compete with the emerging Asian economies for new global markets, the United States must focus on labor productivity across the board, in both low- and high-tech industries. Understanding labor productivity trends and the impacts of working hours on productivity is the first step in this direction.

6.4.2 History of U.S. labor productivity studies

Productivity studies began in the United States in the 1800s and became more common as the Industrial Revolution gained ground by enhancing production through the use of machines. The impact of the Industrial Revolution on U.S. labor must be analyzed with caution. The initial concept of the productivity study appeared at the same time concern was growing about the replacement of human labor by machines.

In 1898, Congress authorized the Commissioner of Labor to investigate and compile the first significant productivity study (Wright 1904). The ensuing report discussed the effects of machine use on labor costs, wages, production costs, productivity, and general standards of living. Qualitative evidence proved that machinery use increased labor productivity. However, obtaining an exact quantitative measurement of the relationship of machine use and increased productivity was difficult.

A comprehensive study of all industrial processes including production labor time and cost was performed. The results were compared with production and labor data from the agricultural sector and industries from different periods. Although the comparisons of data from different industries and other time periods did not provide uniformly conclusive results and inferences, marked decreases in production time and somewhat slower decreases in labor costs were observed.

Since the beginning of industrialization, reductions of production time and labor effort have been the primary goals of manufacturing industries. The reductions infer increases of labor productivity. However, an increase in labor efficiency also indicates that the same production levels can now be achieved by employing fewer people. Between 1900 and the 1930s, the lack of employment was a major concern. Jobs were lost as manufacturing companies reduced labor costs by replacing workers with machines.

Labor productivity studies then focused on the employment issue and showed that the lack of jobs had little impact on productivity. The studies also noted that productivity supported by technology continued to improve. To redistribute productivity gains more evenly among workers while increasing consumer demand for production, shorter working hours were suggested.

Labor productivity studies in the 1940s World War II era focused on resource utilization. The huge labor shortage during the war made it necessary for available labor to be efficient and productivity studies were expanded to cover government and defense industries. A better understanding of government and defense labor productivity was considered vital to the war effort. Labor productivity studies during the Cold War from the 1950s through the 1990s targeted technological contributions to productivity (Klotz et al. 1980). Productivity studies were limited to select industries that generated easily available data.

By the end of the twentieth century, labor productivity was recognized as an important indicator of national progress. The scopes of productivity studies expanded to a number of different industries. The primary focus was on the contribution of labor to economic growth and global competition. Interpreting labor productivity levels of various industrial sectors and comparing them with similar sectors from other countries is fundamental to understanding a nation's performance in the global market and in identifying areas needing improvement. The next section explains how labor productivity is measured by studies.

6.4.3   Labor productivity measures

Labor productivity is defined as the ratio of output produced to the amount of labor input required to produce output. The U.S. Bureau of Labor Statistics (BLS) uses a labor productivity index (U.S. Department of Labor 2011a) calculated as an index of industry output divided by an index of hours:

$$Labor\ Productivity = \frac{Index\ of\ industry\ output}{Index\ of\ hours}\ .$$

where

$$Index\ of\ industry\ output = \frac{Qt}{Qo}$$

and

$$Index\ of\ hours = \frac{Lt}{Lo}.$$

The t indicates the current year and o represents the base year. For a single product industry, it is easy to calculate the output index as the ratio of total number of units produced in the industry in the current year to the base year.

The input hour index is the ratio of total hours worked by all the employees during the current year to the total hours of the base year. This formula can be extended to calculating labor productivity for industries producing multiple products or services. For these industries, output is calculated using the Tornqvist formula:

$$\frac{Q_t}{Q_{t-1}} = \exp\left[\sum_{i=1}^{n} w_{i,t}\left(\ln\left(q_{i,t}\,/\,q_{i,t-1}\right)\right)\right]$$

where

$\dfrac{Q_t}{Q_{t-1}}$ = ratio of output in the current year to previous year.

n = number of products.

$q_{i,t}/q_{i,t-1}$ = ratio of output of ith product in current year to output of previous year.

$w_{i,t}$ = average value share weight for product i.

$$w_{i,t} = (s_{i,t} + s_{i,t-1}) \div 2$$

$$s_{i,t} = p_{i,t}\, q_{i,t} \div \left[\sum_{i=l}^{n} p_{i,t}\, q_{i,t}\right]$$

and $p_{i,t}$ = price of product i at time t.

The Tornqvist formula yields the ratio of output of current year to the previous year (U.S. Department of Labor 2011b). This ratio has to be chained to form a series with the base year:

$$\frac{Q_t}{Q_o} = \frac{Q_3}{Q_o} = \left(\frac{Q_3}{Q_2}\right)\left(\frac{Q_2}{Q_1}\right)\left(\frac{Q_1}{Q_o}\right).$$

This final

$$\frac{Qt}{Qo}$$

is used in calculating the output for labor productivity studies. Other measures such as multifactor productivity evaluate output in relation to multiple inputs, such as capital, labor, and purchases. However, the details of these measures are beyond the scope of this chapter and can be obtained from the U.S. Bureau of Labor Statistics.

Data derived from the concepts of labor productivity measures explained above were used to generate the graphs used in this chapter. The choice of industry output Q in the labor productivity index was the real gross private domestic product (GDP) of the non-farm (primarily manufacturing) sector. The index of hours input L represents labor hours.

GDP refers to the total dollar value of all goods and services provided during a specific period. As an aggregate measure of output, GDP provides the advantage of discounting intermediary changes in goods and services. The graphs used here exclude contributions to GDP from government industries, as it is presumed that the notion of working extensively for profit and resource maximization is lost at times in government industries in favor of overall benefits for society as a whole. It is important that the method and concept used to produce graphs and the input and output parameters involved are understood before we proceed with the ensuing discussion.

6.4.4   *Labor productivity trends*

This section contains three important graphs that depict changes in labor productivity between 1890 and 2000. The smaller the range studied, the more effective the conclusions. Thus, the span between 1890 and 2000 was split into three periods. To determine the trend, a base year was first selected and the productivity output indices of the years under study were represented in comparison to the base year.

The base year to an extent averages price and market fluctuations and other economic changes in the years studied. The base year concept can be thought of as a productivity median approximation. A separate base year was selected and used as a comparator for each of the three graphs. Combining these graphs into a single series creates ambiguity because the productivity representation of each graph is strictly a function of the base year.

Keeping the base year as 100, the productivity output index (GDP index in our case) of each year was given a specific performance value in relation to the base year. These values were plotted on the primary Y axis. The graphs also include a secondary Y axis indicating changes in working

Figure 6.7 Labor productivity and working hours. The graph is plotted showing 1929 as the base year. (*Source:* Kendrick, JW. 1973. National Bureau of Economic Research.)

hours during the years considered. The objective was to understand how working hours changed over the years and their impacts on productivity.

These graphs clearly show how working hours and productivity have changed over the years. Comparing productivity and working hours in parallel reveals the effects of working hour changes on labor productivity.

Figure 6.7 shows the productivity changes from 1889 to 1957. Continued growth in productivity despite the decrease in labor working hours can be observed. The period from 1920 to 1930 reveals a steep decrease in working hours during the Great Depression when shorter hours were established to share work and reduce unemployment. Contrary to general belief, the decrease in working hours had positive effects on productivity as shown in the graph.

In the early twentieth century, the relationship of working hours and industrial productivity remained unknown and businesses tended to continue expending the same labor effort utilized in the labor-intensive pre-industrial era of agriculture. The first few years of the century revealed that decreased working hours exerted no major negative impact on productivity. Rather, productivity increased with decreased hours. This led to the conclusion that the additional hours worked during 60-hour weeks did not significantly improve productivity and represented a substantial waste of human effort. This realization led to a steady decrease in working hours.

Figure 6.8 shows the productivity changes from 1929 through 1970. It is clear that productivity continued to increase over the years. However, in comparison to the previous period (Figure 6.7), the rate of productivity decreased as a result of a relative slowing of the impacts

Figure 6.8 Labor productivity and working hours between 1929 and 1970. Hours gradually stabilized around the 40-hour mark. The graph is plotted showing 1959 as the base year.

of industrialization. The surge in effective utilization of resources that accompanied the advent of industrialization evened out with time.

As previously noted, working hours fluctuated after the war and gradually stabilized around the 40-hour mark. It is important to note that productivity continued to increase during this period despite the stability in working hours. Since the GDP index is calculated per labor hour, this trend clearly indicates the positive effects of development and technology on productivity rather than simply representing an increase in the volume of labor.

Figure 6.9 reinforces the trends observed in Figures 6.7 and 6.8. Productivity gains continued while labor working hours stabilized around the 40-hour mark. Constant exploration of new products and global markets while maintaining high labor productivity is a must for continued economic development of the United States As new technologies and techniques continue to appear, labor must continue to be efficient in coping with the improvements.

Based on Figures 6.7 through 6.9, however, the trend of increased labor productivity, while welcome, may not be convincingly accurate. While the central goal of increasing labor productivity remains unchanged, it is prudent to compare GDP indices of multiple countries to draw conclusions about how U.S. labor performs against labor forces of other countries. The comparison, however, is beyond the scope of this chapter.

It is also interesting to note that the only significant instance where a meaningful comparison between workweeks under 40 hours and labor

Figure 6.9 Increase in labor productivity despite stable 40-hour workweek; 1989 is the base year. (*Source:* U.S. Bureau of Labor Statistics. 2011b. <http://www.bls.gov/opub/hom/>)

productivity was achieved was during the economic depression in the 1930s. The shorter workweeks were dismissed post-depression as "slicing a shrinking pie." The productivity during that period, however, continued to increase. Technically, shorter working hours indicated that more workers shared the work. An increase in the number of workers was expected to decrease labor productivity significantly. For instance, an output of ten units per worker becomes five if two workers are involved. Based on this argument, a productivity increase, despite more workers and fewer hours worked per worker, is suspicious enough to require further investigations favoring shorter working hours.

6.4.5   Impact of other factors on labor productivity

Time is an important aspect of labor productivity. The amount of time a worker works directly impacts how well he or she works. The labor productivity trends discussed thus far explain how changes in working hours affected labor productivity.

The average U.S. worker who has a family and children works about 8.6 hours per day. His or her work efficiency is not just a function of working hours; it also depends on the number of hours spent on activities other than work (sleeping, leisure, sports, and family activities). Each segment shown in Figure 6.10 has a significant impact on the productivity of a worker. Finding a balance among these activities is important in increasing overall labor productivity and enhancing the quality of life.

On average, a working person needs 6 to 8 hours of quality sleep. If the working hours per day increased and compromised sleep, a worker

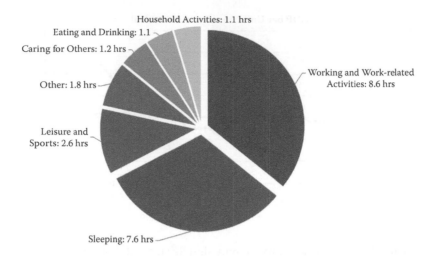

Figure 6.10 Time use on average workday, for employees aged 25 to 54 with families and children in 2010. (*Source:* U.S. Department of Labor. 2011c. <http://www.bls.gov/news.release/archives/atus_06222011.pdf>)

would more likely be inefficient and contribute to negative productivity caused by quality problems, accidents, and illnesses. Similarly, longer working hours cause a worker to spend less time with his or family. This may lead to disturbances in family life that create distractions during work hours and ultimately decrease productivity. The effects of working hours are also functions of the nature of the job performed. A worker's contribution to his job depends on his physical and psychological states.

On the physical front, a worker coordinates his body and mind to complete a task. Effectiveness of both the body and mind work in conjunction with nature of the job and its capacity to stimulate. A human body can expend only a limited amount of mind and body activity at work in relation to the type of task performed. Taxing a worker's capacity beyond physical limits will affect the quality of his or her output. Fatigue, boredom, lack of vigilance, stress, and accidents result from pushing workers beyond their physical and mental limits. A subtle balance between work and rest must be maintained for a worker to be productive. This balance is maintained through voluntary and involuntary rests at work. Voluntary rests are prescribed pauses such as coffee and lunch breaks. Involuntary rests include spontaneous pauses and unscheduled pauses between jobs such as switching to routine tasks or reading news on the Internet.

From a psychological perspective, Parkinson's law, "Work expands so as to fill the time available for its completion," usually holds true. In other words, the longer the working hours, the greater is the tendency of a worker to be inefficient. A worker on a 6-hour per day schedule will work efficiently to meet his deadline utilizing what is called deadline psychology. On the other hand, a worker on a 10-hour per day schedule assumes a subconscious comfort zone and believes he has more time than he actually does, then fritters away the comfort zone time being non-productive.

This argument, however, is entirely subjective and depends on the nature of job, supervision, and other factors. The state of Utah is a notable exception. It has achieved reasonable success by declaring 4-day, 40-hour workweeks for all public sector institutions.

The concept of a stable 40-hour week across different kinds of jobs demanding different skill sets has an inherent flaw that must be addressed. Increasing the workday beyond the 8-hour mark has proven wasteful for the reasons stated above. The concept of working 6 hours a day is still not widely accepted. The increased employment opportunities and improved labor effectiveness based on decreased working hours have not proven conclusive.

In summary, from an environmental view, decreasing labor-working hours is preferable as it implies a decrease in the use of natural resources. Also, the concept of having more time for personal life removes the dependence on consumer products solely associated with saving time for work, for example, fast foods and electric appliances. This is likely to change our consumption dynamics and aid environment sustainability. From a productivity standpoint, it is still not proven that decreasing working hours decreases productivity. From a worker's view, fewer working hours mean improved quality of life, more time for family, more time for sleep and leisure, more involvement in community and social activities, increased employment opportunities, and more equitable income distribution.

The shorter working hours imply a decrease in wages and workers are more likely to resist the change or counter it by working overtime—thus defeating the purpose. Shorter working hours mean hiring more workers. An employer is likely to resist hiring more workers unless the result is a significant increase in productivity. Increasing the number of workers means increasing costs for recruiting, hiring, insurance, training, and development. Implementation of shorter working hours, despite its obvious advantages, has not become standard for the reasons cited above.

6.5 Recent developments and conclusions

In the twenty-first century, the U.S. workforce became dominated by service industries and information technology. The decline in primary occupations such as farming and secondary occupations like manufacturing must be analyzed with caution. Less than 30% of the modern U.S. workforce is currently employed in manufacturing (Figure 6.11). This shift away from manufacturing is likely to affect economic development in the long run. Tertiary occupations like service industries offer valuable services to enhance productivity but are not directly involved in creating

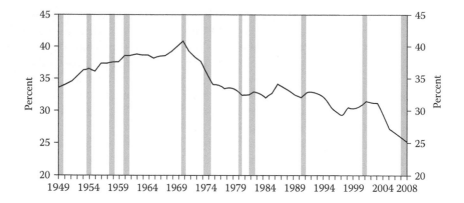

Figure 6.11 Percentage of workforce in the manufacturing sector since 1949. Note decrease in manufacturing workforce in the twenty-first century. The shaded region indicates slow economic progress. (Source: Fleck, S. et al. 2011. *U.S. Department of Labor Bureau of Labor Statistics Monthly Labor Review.*)

goods. Although services are very important to national development, neglecting primary and secondary occupations is very likely to transform the United States from a "creating" economy to a "dependent" economy.

Shorter working hours and working weeks as possible solutions to improve labor productivity need serious consideration. Labor productivity in the manufacturing sector has slowed since the beginning of the twenty-first century (Figure 6.12) mainly because of offshoring of manufacturing jobs to countries with low labor costs. For the U.S. manufacturing sector to sustain and compete globally, labor costs must decrease and productivity must increase. This is possible only if manufacturing industries adopt

Labor Supply and Factors Affecting Productivity							
Category	Levels				Average Annual Rate of Change		
	1980	1990	2000	2010	1980–90	1990–2000	2000–10
Labor supply (in millions, unless noted):							
Total population......................	228.0	250.3	275.7	300.3	0.9	1.0	0.9
Population aged 16 and older..........	172.7	192.8	213.1	236.7	1.1	1.0	1.1
Civilian labor force...................	107.0	125.9	140.9	157.7	1.6	1.1	1.1
Civilian household employment........	99.3	118.8	135.2	151.4	1.8	1.3	1.1
Nonfarm wage and salary employment ...	90.4	109.4	131.8	152.0	1.9	1.9	1.4
Unemployment rate (percent)..........	7.2	5.6	4.0	4.0	−2.4	−3.3	−.1
Productivity:							
Nonfarm labor productivity (1992 = 100)	82.00	95.28	116.23	153.54	1.5	2.0	2.8

Sources: Historical data, Bureau of Economic Analysis, Bureau of Labor Statistics; projected data, Bureau of Labor Statistics.

Figure 6.12 Labor productivity in 2010 and average annual rate of change in productivity from 2000 to 2010.

conscious efforts to improve productivity instead of choosing the easy option of offshoring labor to other countries to increase profits.

References

Atack, J. and Bateman, F. 1992. How long was the work day in 1880? NBER Historical Working Paper 15. *Journal of Economic History*, 21(1), 129–160.

Carter, S.B., Gartner, S.S., Haines, M.R. et al. 2011. *Historical Statistics of United States*, Millennial Edition Online. Cambridge: Cambridge University Press.

Dimitri, C., Effland, A., and Conklin, N. 2005. *History of American Agriculture: The 20th Century Transformation of U.S. Agriculture and Farm Policy*, Economic Information Bulletin 3. Washington: U.S. Department of Agriculture Economic Research Service.

Fleck, S., Glaser, J., and Sprague, S. 2011. The compensation–productivity gap: a visual essay. *U.S. Department of Labor, Bureau of Labor Statistics Monthly Labor Review*, January.

Giampietro, M., Bukkens, S.G.F., and Pimentel, D. 1993. Labor productivity: a biophysical definition and assessment. *Human Ecology*, 21, 229–260.

Holt, S.J. 1970. Impact of the industrialization of the hired farm work force upon the agricultural economy: changing problem of the agricultural work force. *American Journal of Agricultural Economics*, 52, 780–786.

Hunnicutt, K.B. 1984. The end of shorter hours. *Labor History*, 25, 373–404.

Jones, E.B. 1975. State legislation and hours of work in manufacturing. *Southern Economic Journal*, 41, 602–612.

Kelly, M. Overview of the Industrial Revolution: The United States and industrial revolution in the 19th century. http://americanhistory.about.com/od/industrialrev/a/indrevoverview.htm accessed December 8, 2011.

Kendrick, J.W. 1973. *Postwar Productivity Trends in the United States*. Washington: National Bureau of Economic Research.

Kendrick, J.W. 1956. *Productivity Trends: Capital and Labor*. Washington: National Bureau of Economic Research, pp. 3–23.

Klotz, B., Madoo, R., and Hansen, R. 1980. Study of high and low labor productivity establishments in U.S. manufacturing. In: *New Developments in Productivity Measures*, J.W. Kendrick and B.N. Vaccara (eds.). University of Chicago Press, 239–292.

Rones, P.L, Randy, E.I., and Gardner, J.M. 1997. Trends in hours of work since the mid-1970s. *Monthly Labor Review*, (April) 3–14.

U.S. Census Bureau. 2012. *Statistical Abstract of the U.S., 2012*, 131st ed. Section 12: Labor Force, Employment, and Earnings. <http://www.census.gov/compendia/statab/>

U.S. Census Bureau. 2011. *Statistical Abstract of the U.S., 1950–2011*. <http://www.census.gov/prod/www/abs/statab/1951-1994.htm>

U.S. Census Bureau. 1950. *Special Report: Agriculture. A Graphic Summary*, Vol. V, Part 6, pp. 69–73.

U.S. Department of Labor, Bureau of Labor Statistics. 2011a. *Industries at a Glance: Manufacturing NAICS*. pp. 31–33.

U.S. Department of Labor, Bureau of Labor Statistics. 2011b. *Industry Productivity Measures: Handbook of Methods*. Chap. 11. <http://www.bls.gov/opub/hom/>

U.S. Department of Labor, Bureau of Labor Statistics. 2011c. *American Time Use Survey: 2010 Results.* <http://www.bls.gov/news.release/archives/atus_06222011.pdf>

Whaples, R. Hours of work in U.S history. http://eh.net/encyclopedia/article/whaples.work.hours.us. Accessed December 13, 2011.

Wright, C.D. 1904. Hand and machine labor. In *13th Annual Report of the Commissioner of Labor, 1898–1899.* Washington: U.S. Government Printing Office.

chapter seven

Comparison of productivities of manufacturing workers in U.S. and selected developed and fast developing economies

Vignesh Ravindran, Aashi Mital, and Anil Mital

Contents

7.1   Introduction

The twenty-first century is witnessing a gradual but definite shift in the epicenter of global manufacturing—from traditionally *developed economies* of the twentieth century to modern *emerging economies*. The purpose of this chapter is to analyze this geographical shift in manufacturing.

Considering that global change cannot be attributed easily to a single event over time, it is necessary to review the conditions in both developed and developing economies over the past 50 years, and in some cases, even longer, in an attempt to understand the causes and conditions leading to this change. Regardless of the nature of these changes—whether caused by technology advances, market conditions, manufacturing policies, or labor scenario—the purpose is to understand them and explore how they affected worker productivity in the manufacturing sector.

The analysis and comparison, however, are limited to a select number of developed and developing economies. It is expected that such reflections will provide us with a deeper and more extensive understanding of the current changes in manufacturing industries and that the actions and effects that led to the current state will help dictate future actions.

Industrialization is the backbone of a country's economy. Industrialized nations (also known as developed countries) have seen economic prosperity over the past 100 years. They have effectively utilized manufacturing technology to increase their gross domestic products (GDPs) and thus maintain high standards of living.

Western European countries such as the United Kingdom, France, Germany, and Italy have enjoyed the benefits of industrialization and economic dominance. Early in the twentieth century, the United States emerged as a powerful economic engine on the global scene. Understanding modern manufacturing and investing heavily in technological research and development provided the United States the opportunity to overtake Europe as the most developed economy in the twentieth century.

The second half of the century gave birth to a resurgent Japanese manufacturing sector that gained strength due to its innovative manufacturing practices. Japan overtook the United States and Europe as the world's fastest growing economy. Western Europe, the United States, and Japan are classed as developed economies.

Over time, many other nations began to make rapid strides toward industrialization while strengthening their economies in the process. Four states currently have the largest and fastest growing economies: Brazil, Russia, India, and China. These nations, along with South Africa, form a unique group of emerging advanced economies known as *BRICS*. The general characteristics of all these countries except Russia include: (1) late onset of industrialization, (2) fast growing GDPs, and (3) strong influence on global manufacturing. BRICS nations fall into the developing economies category.

Since the primary goal is to explain the position of the U.S. manufacturing sector among other developed and developing economies, the United States will be compared with other countries on the basis of manufacturing productivity.

Section 7.2 discusses developed economies. The analysis covers changes in the manufacturing sector, traces productivity changes in Japan and certain Western European countries, and compares results to the U.S. manufacturing sector. This chapter also includes a brief overview of the manufacturing sectors of developing economies. Conditions that led to the industrialization of BRICS and changes in their manufacturing industries over the years are discussed. Various indicators are used to compare their performance in detail. The last section compares the manufacturing productivities of Japan, the United States, Europe, and BRICS.

7.2   Developed economies

7.2.1   Japanese industrialization overview

By the end of the nineteenth century, when the world economy was dominated by European colonial empires, Japan succeeded as the first non-Western country to be industrialized on a par with its Western counterparts. The Japanese had sociocultural traditions that strongly favored industrialization. Its inherent interest in manufacturing (known as *monotsukuri* or "making things") is evidenced by its rapid rise as a developed economy.

The Japanese feudal system throughout the nineteenth century laid the early foundation for an industrial Japan. Japan then had a closed economy, remained isolated, and believed in self-sustenance (Gordon 2003). This led to the strong development of communications and transportation networks within Japan and a national integration of its economy.

The development of a strong endogenous structure before industrialization uniquely favored Japan (Ohno 2000). Today's emerging nations such as India and China are forced to reach the same levels of industrialization over a short time and hence cannot follow the ideal sequence of developing a structure that allows industrialization to flourish.

The advent of industrialization in Japan is largely attributed to a political revolution known as the Meiji Restoration of the Imperial Rule that ended the feudal system in 1868. The primary objective of the Meiji regime was to maintain national sovereignty in the face of colonial threats. To prevent European colonization, the regime planned to combine Western technology with Eastern traditions to transform Japan into a powerful industrial nation.

Subsequently Meiji Japan adopted a number of Western policies such as a market economy, privatization, and enterprise capitalism. The Meiji theme of *shokusan kougyou* (industrialization drive) was embraced by all quarters of Japanese society. Early in the transition phase, the government set up prototype factories based on Western models and hired foreign experts to manage them.

After Japanese entrepreneurs mastered Western technology and machinery, the Meiji regime privatized a number of government enterprises and supported them with favorable policies and investments. These private enterprises chose appropriate Western technology and developed their businesses to suit local Japanese needs. The formation of *zaibatsu* (large family-owned corporations) also played a major role in Japanese industrialization. Sharing of information, technology, and financing became easier under a centralized system (Shoten 1989).

On the political front, Japan believed in accumulating wealth and power, and industrialization was the key. The growing Japanese military organization protected the country's economic interests in the Eastern Pacific and shielded Japanese enterprises from European competition. This combined effort of government policies to support industrialization and the role of dynamic private entrepreneurship to embrace the policies and adopt them to achieve radical industrial progress transformed Japan into a dominant world force.

The Japanese model of industrialization is difficult to replicate in a modern environment. The concept of military expansion that aided Japanese industrialization is no longer a viable option. The developing Japan of the 1900s had the luxury of choosing suitable policies regardless of their impacts on the international community. Unfortunately, today's developing nations must follow international rules and policies that slow their economic progress.

Japan believed in using Western ideas to develop indigenous technology and resources. Today's developing countries have to achieve fast growth and face competition in a global market. They are forced into over-reliance on foreign direct investments and technology without sufficient time to develop their own finances and inventions—a necessary compromise that will haunt them over the long run.

7.2.2 Comparison of productivity changes in Japan and the United States

1930–1950: In 1930, when U.S. industries were struggling with economic depression and overproduction, the far smaller Japanese economy was thriving at an average growth rate of 5%. Japanese industries made rapid advancements in engineering and concentrated most of their efforts on manufactured goods. Japan became a major importer of raw materials and major exporter of manufactured goods.

Between 1930 and the end of World War II, the Japanese index for consumption goods rose from 100 to 154. The index for investment goods (machine tools, plant equipment) rose from 100 to 264 (Dower 1990). Manufacturing production indices showed an increase of 24% and the nation's manufacturing labor force increased to 8 million. Nearly 70% of the industrial workforce was employed in heavy manufacturing industries by the 1940s.

World War II shifted the momentum back to the United States. Japan nearly lost a quarter of its wealth because of the war. Most industries and industrial towns were bombarded during the war. The rise of Japanese manufacturing industry and the growth of the Japanese economy after World War II is the primary interest here. In the aftermath of the war, the economic rebuilding of Japan revolved around the concept of promoting manufactured goods. Post-war Japan completely reformulated its industrial policies. Strong investments in education and physical infrastructure were made to sustain industrialization over the long run (Okada 1999). The support of Japan's Ministry of International Trade and Industry (MITI) in industrial growth was crucial to the nation's success.

1950–1990: In 1950, Japan's manufacturing output per hour index was a mere 3.9 in comparison to 19.5 in the United States (U.S. Bureau of Labor Statistics 2011a). As a U.S. ally during the Cold War, Japan enjoyed the advantages of access to American manufacturing technology and skill at relatively low cost. Japan benefited from the open market system that spread worldwide after World War II. MITI successfully negotiated access to foreign technology in return for opening Japanese markets to foreign products (Mosk 2001).

Japanese industry invested heavily in developing modern manufacturing techniques and practices. Elimination of production waste, invention of the value addition model, process improvements, inventory management techniques, and development of Just-in-Time (JIT) and Total Quality Management (TQM) programs are some of the advanced practices that aided Japan's rise as the world capital of manufacturing. By utilizing labor policies like the lifetime employment system and seniority (*nenko*) wage system, Japanese manufacturers of the 1960s managed an ideal harmony between labor and technology (Hitomi 1993).

Between 1950 and 1970, the Japanese manufacturing index grew more than tenfold from 100 in 1950 to nearly 1350 by the end of 1970s (Johnson 1982). Japanese manufacturing output per hour exceeded that of the United States by 1970. Japanese manufactured goods dominated the world market for almost 20 years until 1990 (Figure 7.1).

1990–2000: The "Lost Decade" resulted from the collapse of the Japanese asset price bubble. Japanese manufacturing declined sharply due to a nationwide economic crash caused by real estate and stock price inflation. Decline in demand for manufacturing goods, especially machine

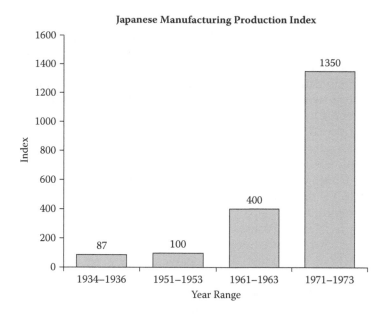

Figure 7.1 Japanese Manufacturing Production Index. (*Source:* Johnson, C.A. 1982. *MITI and the Japanese Miracle: The Growth of Industrial Policy 1925–1975.* Palo Alto, CA: Stanford University Press. With permission.)

tools, in the post-bubble era affected Japanese manufacturing to a huge extent. While Japanese industry attempted to reinvent the manufacturing sector after the crash, U.S. competitors were gaining speed.

The real gross domestic product (GDP) per hour worked shown in Figure 7.2 compares the exponential Japanese progress in comparison to that of the United States between 1980 and 1990. Between 1990 and 1995, however, Japan's GDP dropped significantly as a result of the economic crash. From 1995 to 2000, Japan made recovery plans but its GDP remained stagnant in contrast to the growing U.S. GDP. Between 2000 and 2010, both nations exhibited overall decreases in GDP growth although the United States fared better.

Figure 7.3 shows GDP per person employed. Comparing Figures 7.2 and 7.3, Japan's GDP per hour worked and GDP per person employed show similar decreasing trends. From 1990 to 1995, GDP per person employed declined sharply in comparison to GDP per hour worked. This reflects changes in labor conditions in Japan during the crash. Industries switched to hiring temporary workers and the number of hours worked per employed person decreased. In the manufacturing sector, industries started automating their assembly lines to decrease labor costs and full-time workers were sacrificed for part-time workers hired by the hour based on production demand.

Real GDP per Hour Worked

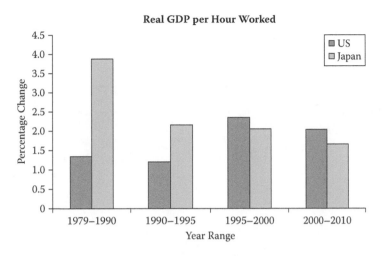

Figure 7.2 Comparison of United States and Japanese real GDP per hour worked. (*Source:* U.S. Bureau of Labor Statistics. 2011a. International Comparisons of Productivity and Unit Labor Cost Trends: 2010 Data Tables.)

Real GDP per Employed Person

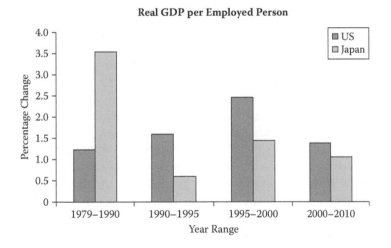

Figure 7.3 Comparison of United States and Japanese real GDP per person employed. (*Source:* U.S. Bureau of Labor Statistics. 2011a. International Comparisons of Productivity and Unit Labor Cost Trends: 2010 Data Tables.)

2000–2010: Since the economic crash, the scarcity of natural resources in Japan is growing (Figure 7.4). The resulting increases in costs of raw materials led to high manufacturing costs (Deloitte 2010). Internal competition caused many Japanese companies producing similar products to lose significant revenue through pricing advantages. Japanese overseas pricing strategy has been criticized for compromising profit maximization

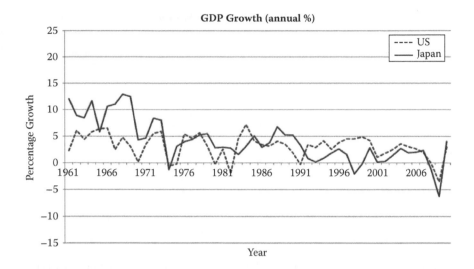

Figure 7.4 Comparison of United States and Japanese GDP growth rate. (*Source:* U.S. Bureau of Labor Statistics. 2011a. International Comparisons of Productivity and Unit Labor Cost Trends: 2010 Data Tables.)

for an increase in market share (dumping), resulting in a decrease in value added per worker. The rise of competitive manufacturing sectors in South Korea, Singapore, and Taiwan combined with their low labor costs also stunted Japanese growth in Asian markets.

Expensive domestic labor cost and a lack of young workers have become major concerns. To counter these issues, Japanese manufacturing industries embraced factory automation and assembly line robotics. Automating an entire manufacturing plant decreases reliance on expensive labor but requires huge capital investments. An increase in capital investment per unit labor is believed to increase labor productivity in the long run.

As in the United States, the pressures of increasing domestic labor costs forced Japanese industries to outsource their manufacturing jobs to countries with low labor costs, like China, to remain competitive. The Japanese overseas production percentage increased from 8% in 1994 to nearly 17% in 2008. This trend of outsourcing manufacturing jobs resulted in a decrease in manufacturing development activities, causing an ongoing migration of existing manufacturing technology to other countries. The number of local manufacturing jobs in Japan has declined continuously from 149.6 million in 1994 to 107.3 million in 2008.

2011: Japan identified innovation as the core strength of its manufacturing industries. It targets products at the high end of the technological spectrum (carbon fibers, fine chemicals, high-end electronics, next-generation vehicles, and robotics) to avoid competition from low labor cost countries and sustain its manufacturing sector (Japan Ministry of Trade and Industry 2010). Saturated domestic markets prompted Japan to explore and capture consumer markets in emerging countries like India and China.

Japan recognizes software development and information technology (IT) integration in manufacturing as key areas needing improvement. Another concern is the growing trend of worker migration toward tertiary jobs and service sectors. Manufacturing jobs have become less attractive in Japan. Other challenges include the energy crisis after the Fukushima nuclear disaster on March 11, 2011 that led to a nationwide shutdown of nuclear power plants. The ensuing energy shortfall and subsequent government order to reduce industrial electrical consumption by 15% forced more outsourcing of manufacturing jobs. How the Japanese manufacturing sector of the twenty-first century will survive the intense global competition from emerging economies remains to be seen.

7.2.3   European industrialization overview

Europe dominated the manufacturing sector for more than 200 years, since the beginning of the Industrial Revolution in England in the early nineteenth century. European nations are considered developed countries although they lost significant manufacturing momentum after two world wars. The rise of the U.S. manufacturing sector after World War II and the huge expansion of the Japanese manufacturing sector during the 1970s effectively ended the domination of European goods in the global market.

Although European countries continue to play a major role in world manufacturing, much of their technology has been found wanting against their American and Japanese counterparts. Because of increasing domestic labor costs, European countries, like other developed nations, continue to embrace the trend of offshoring labor-intensive manufacturing jobs to China and other countries with low labor costs. As the GDP growth rates of emerging Asian economies continued to overtake those of the developed economies, the decline of manufacturing (de-industrialization) in Europe that started in the 1970s continued. The European manufacturing sector is briefly discussed here.

Figure 7.5 shows the 2008 manufacturing contributions of the European Union (EU) member states. Of the 27 members; Germany, France, Italy and the U.K. led the manufacturing sector over the years. For this reason, the discussions and comparisons of European conditions focus on progress in these four countries. England was the first of the four to industrialize (European Commission 2012). The Industrial Revolution started there at the end of the eighteenth century when a series of inventions changed various industries, particularly textile production. The development of the first machine tool in England is widely regarded as the start of the modern manufacturing era. France followed suit in industrialization by the early

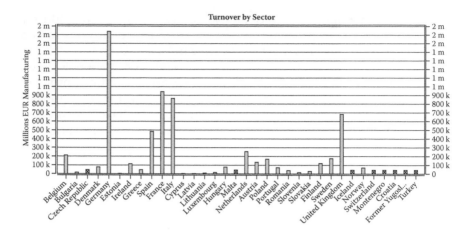

Figure 7.5 Manufacturing turnover of members of European Union, and seven other countries, 2008. Germany, France, Italy, and the United Kingdom are largest contributors in the manufacturing sector (*Source:* European Commission 2012 Eurostat: European Statistics.)

nineteenth century. Germany arrived late on the international stage by the turn of the nineteenth century. Italy was the last to make the transition.

7.2.4   Comparison of changes in European and U.S. manufacturing productivities

1930–1950: The European manufacturing sector was largely built around the policy of colonization around the world. Colonization presented a number of advantages: cheap and readily available raw materials, inexpensive labor, open markets with price controls, and tax subsidies. Europe, however, considered manufacturing more as a means to achieve capital gains than as a separate economic component. It considered manufacturing an adjunct to business strategies (Ward 1994). Hence, European manufacturing, despite its dominance in the nineteenth century, failed to develop proper structure and organization. Its ongoing dependence on colonial markets obviated the need to develop the commercial and sales aspects of production and hampered post-war progress throughout the continent.

By the early twentieth century, investments in developing manufacturing technology significantly decreased because finances were directed more toward colonial maintenance and later dedicated to wars and defense. Europe suffered major setbacks because of heavy involvement in two world wars. Most European colonial markets were lost by the end of World War II and European economies including the manufacturing sectors were crippled completely.

Conversely, U.S. manufacturing advanced by the 1900s. Technological efforts focused on developing new techniques for developing new products and also for improving various aspects of production, assembly, and distribution. Production advances such as assembly lines, mass production, and use of replaceable parts gained huge momentum. In terms of investment, limited participation in world wars and the economic advantage provided by the situation in Europe allowed U.S. manufacturing firms to invest heavily in manufacturing technology.

European markets presented little potential because they were already saturated. European local markets were small, diverse, and lacked homogeneity. In comparison, American markets were large, able to grow, and relatively homogeneous. A number of other geopolitical factors also combined to favor American manufacturing. By the end of 1950s, the United States replaced Europe as the most dominant force in global manufacturing.

1950–1970: The economic recovery of Europe after World War II was accompanied by a major overhaul of industrial policies, organizations, labor, infrastructure, and sales. Like Japan, Europe (especially the western half) benefitted from an alliance with the United States during the Cold War. The rise of European manufacturing proceeded hand in hand with the rise of American manufacturing in the 1950s. America shared its manufacturing technology and provided massive financial aid to rebuild Europe in an attempt to thwart Soviet influence through the European Recovery Program, also known as the Marshall Plan.

Between 1950 and 1970, the European manufacturing sector discovered modern techniques and re-entered the global market as a manufacturing force. This period was marked by a rapid increase in manufacturing productivity across most of Europe and massive growth in the manufacturing shares of GDPs (Crafts and Mills 2005).

1970–2000: The rapid rise of European manufacturing sector and the economic boom starting in the 1950s gradually came to a halt by the 1970s (Figures 7.6 through 7.9). Significant changes in international economic policies including the end of the gold standard in 1971 and the 1973 oil crisis resulted in a major stock exchange crash that severely impacted European economies (Hesse and Tarkka 1986).

Energy for industrial production became expensive and Europe concentrated on finding alternate energy sources. Since then European manufacturing has fallen behind the levels of the United States and Japan. While Japan continued to invest in manufacturing research and development (R&D) to remain competitive and pursued process innovation and quality, European efforts to sustain the manufacturing boom despite investments in R&D were insufficient, and the countries lost their technological edge.

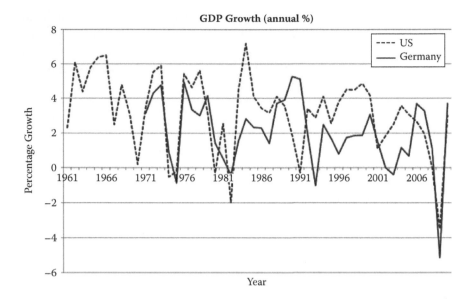

Figure 7.6 GDP growth comparison of Germany and United States. (*Source:* World Bank 2012. World Development Indicators.)

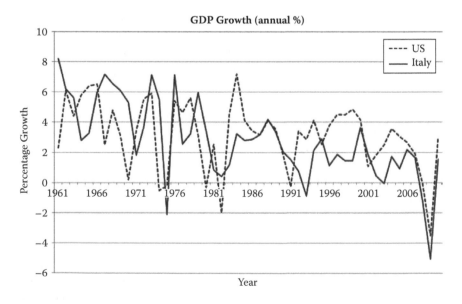

Figure 7.7 GDP growth comparison of Italy and United States. (*Source:* World Bank 2012. World Development Indicators.)

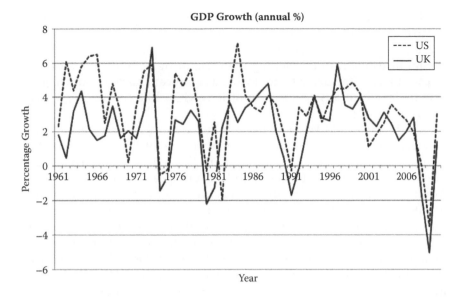

Figure 7.8 GDP growth comparison of UK and United States. (*Source:* World Bank 2012 World Development Indicators.)

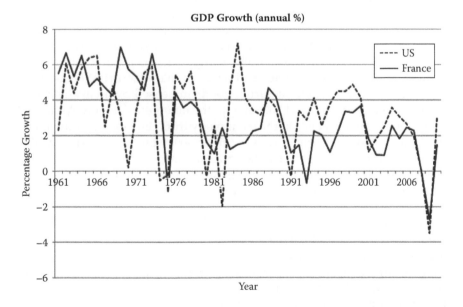

Figure 7.9 GDP growth comparison of France and United States. (*Source:* World Bank 2012. World Development Indicators.)

By the 1980s, European labor composition changed visibly as manufacturing jobs were steadily lost to the business and service sectors. As the service sector contributions to GDPs continued to increase, lack of major capital investment in developing manufacturing technology caused the European manufacturing sector to lose a significant market share to Japan and the newly industrialized countries (NICs) like Singapore, Taiwan, and Korea.

By the post-Cold War era, Eastern Europe was liberated and opened new markets to the European manufacturing sector. The formation of the European Union removed trade barriers through a single market program. Labor, market, monetary, and governance policies standardized throughout Europe. European manufacturing witnessed a sudden surge in internal sales. The worldwide globalization created new markets for European products in Asia, Russia, and South America. Between 1985 and 1995, Europe saw a 30% increase in total production and around 45% of total production was exported. Nearly 63% of manufacturing exports went to fellow European countries and 37% went elsewhere. Despite increases in labor and energy costs, European manufactured goods continued to dominate the world market because of quality.

Between 1988 and 2000, an additional 7% increase in manufacturing exports occurred due to globalization. However, despite the increase in sales, the number of people working in manufacturing gradually decreased at nearly 1% per annum (24 million in 1985 to nearly 22 million in 2000), as the service sector percentage of employment increased.

2000–2010: This period witnessed the rise of newly emerging economies such as China, South Korea, Brazil, and India. They continue to play a significant role in the manufacturing sector as Europe continues to compete with the United States and Japan in manufacturing technology. European exports of manufacturing goods increased by an average 4.5% annually between 2000 and 2008 due to the emergence of new Asian and Latin American markets. On the technology front, Europe in parallel with the United States performed well in science-based industries such as fine chemicals, mechatronics, aerospace engineering, e-manufacturing, computers, and electronics. European market shares in automobiles, chemicals, and pharmaceuticals have grown significantly between 1996 and 2006 (Johansson 2008).

In the labor area, aging and cultural diversity of working populations are proving difficult challenges for the European manufacturing sector. Increased domestic labor costs led to offshoring of labor-intensive low-tech consumer products to emerging Asian countries with lower labor costs. Offshoring non-core manufacturing jobs created an advantage. European production became energy efficient and increased labor productivity by nearly 45% between 1995 and 2005. On the downside, although the strength of European manufacturing over Asian manufacturing is superior technology and product innovation, offshoring production activities to Asian countries with weak intellectual property protection laws led to technological leaks and imitation products that caused significant losses of revenue (Fagerberg et al. 1999).

Despite the decrease in direct contributions of manufacturing to the GDP (less than 20%), manufacturing continues to play an important role in the GDP as services directly or indirectly related to manufacturing contribute to an additional 30% to the total GDP. As the role of the business and service sector continues to grow in Europe, manufacturing jobs lost their attractiveness and labor continues to move toward the service sector. This drop in manufacturing jobs is a major concern because the sustenance of the tertiary sector in Europe depends heavily on manufacturing.

Limited energy and natural resources have caused Europe to lose its advantage in capital- and resource-driven industries such as steel and fiber production. Nuclear power as an alternate energy source has been reluctantly welcomed amid safety concerns as increasing energy costs continue to be problematic. Growing environmental concerns arising from production and consumption activities demand improved manufacturing techniques and product innovation. Sustainable manufacturing now draws more attention in terms of investment and resources as European manufacturing faces a difficult twenty-first century (European Commission 2003).

2011: The rise of new dynamic markets in emerging economies has led to increased demand for regionally customized products. Maintaining cost competitiveness in such a market environment demands efficient supply chain strategies. The European manufacturing sector must improve its supply chain strategy to cater to customized needs of emerging economies. Process innovation and expansion of sales and distribution are other areas requiring improvement in the manufacturing sector.

The European manufacturing sector has experienced a continuous deterioration of labor environments as manufacturing jobs continue to shift to low-cost emerging economies like China and Taiwan. The biggest challenge for Europe is to prevent manufacturing from migrating because that sector is the backbone of the European economy. The economic crash of 2008 highlighted the value of highly skilled labor in the manufacturing sector. Hiring highly skilled labor on a temporary basis mitigated the negative impact of the crash on manufacturing industries. Europe must maintain a significant pool of highly skilled manufacturing labor for the future. Establishing a secure supply of raw materials is another concern because Europe continues to depend on foreign imports for raw materials in key manufacturing industries.

European manufacturing, like manufacturing in Japan and the United States, identifies technology as its core strength. Europe targets areas of modern manufacturing, including nanotechnology, hybrid materials, new polymers and catalysts, and fine chemicals. Lack of major financial investment in manufacturing firms has been a major challenge that has exerted a negative impact on the global competitiveness of European manufacturing. Encouraging investments in manufacturing is vital for future progress.

The environmental impacts of manufacturing and production are also growing concerns for the future as European manufacturing concentrates on product and process innovation in an attempt to achieve sustainable, environmentally friendly manufacturing.

7.3   Developing economies

China, Brazil, Russia, India, and South Africa are five of the largest emerging economies of the twenty-first century. Together they are beginning to challenge traditional manufacturing giants in the global market. The rise of manufacturing sectors in these countries and their roles in global manufacturing are discussed in this section.

7.3.1   Rise of Chinese manufacturing sector

China started her post-war recovery by following the Soviet economic model based on centralized economic planning and major economic decisions in the hands of governments. The first signs of Chinese industrialization started in the mid-1950s with Soviet investments in manufacturing plants and machinery. Chinese industrial production increased by nearly 20% then. Between 1960 and 1970, the Soviet model of government-controlled industrial development proved inefficient and a slowdown in economic progress prompted major modifications of the model including decentralization of authority to local administrative bodies.

By the end of 1970, Soviet influence on China receded as Western Europe and Japan started supporting Chinese industrial recovery. Despite the temporary surge in production and small growth in economy, the late 1970s saw a major decline in production (by nearly 15%) due to political disruptions affecting industrial performance across the country. Between 1980 and 1990, China finally achieved political stability and gradually emerged as a strong economic force as it embraced an open market system and foreign trade. By the end of 1990, foreign trade was an essential part of Chinese manufacturing. Japan, the United States, and West Germany were its major trading partners. Between 1990 and 2000, China achieved consistent growth in GDP and its manufacturing sector was a significant contributor (Figure 7.10). By 2001, China joined the World Trade Organization and opened its manufacturing sector to global markets.

Twenty-first century China is growing into one of the most powerful manufacturing bases in the world. Its manufacturing sector has a number of advantages over U.S. and European sectors. Since China was late to

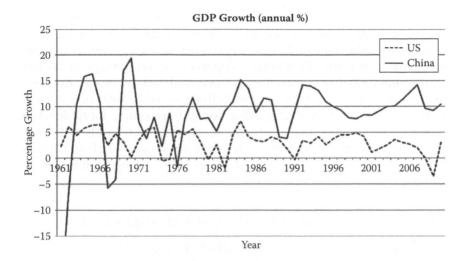

Figure 7.10 GDP growth comparison of China and the United States. (*Source:* World Bank 2012. World Development Indicators.)

embrace large-scale industrialization, its markets are relatively young and have massive potential. As the Chinese economy continues to improve, so does the standard of living of its people and this increases local demands for manufactured goods. China has a healthy and vibrant business environment that allows its manufacturing sector to thrive. China has abundant natural resources and this ready supply of raw material is a big boost to its manufacturing sector.

From 2000 to 2010, Chinese manufacturers exploited the strength of the country's inexpensive labor force (Banister and Cook 2011). They produced acceptable quality goods at low cost that could compete well against U.S., Japanese, and European goods in the global market. China remains the most preferred destination for foreign outsourced production and direct investments due to the availability of semi-skilled and inexpensive labor, improving manufacturing infrastructure, and healthy domestic demand (Zheng 2005). Segments of European, U.S., and Japanese manufacturing have migrated to China as China continues its rise as the center of global manufacturing.

However, despite low cost labor and migration of manufacturing jobs, Chinese manufacturing remains at the low value end of the supply chain (Deloitte 2011). Manufacturing, assembly, and production without the support of core indigenous technology undervalue the Chinese manufacturing effort. Technology is an area where China can improve. Although China continues to rely on foreign technology without major research and development activity of its own, its government has taken positive steps to change this scenario.

Chinese manufacturing faces a long list of modern challenges. Lack of proper organization and structure, poor business and management aspects of production, underdeveloped sales and marketing strategies, poor protection against intellectual property theft, minimal process innovation, suboptimal operational excellence, and lack of proper domestic and international market development efforts are some areas of challenge. Abundant resources, a strong labor force, burgeoning local demand, growing foreign outsourced production, expanding research and development, and improved manufacturing infrastructure will allow Chinese industries to grow into the twenty-first century's strongest manufacturing base.

7.3.2   Industrialization in Brazil, India, and South Africa

7.3.2.1   Changes in Brazilian manufacturing sector

Industrialization in Brazil started in the 1930s, but the country played no role in mainstream manufacturing until the 1950s when it struggled to compete against superior products of established foreign manufacturers. The government decided to promote local manufacturing through a series of strict exchange control and import tariff policies. The general objective was to replace foreign consumer and capital goods with locally manufactured equivalents. This policy of substituting foreign products with domestic goods is known as import substitution industrialization. As a result of this policy, Brazilian manufacturing developed into a large and diverse sector producing a wide variety of products. Between 1965 and 1970, Brazilian manufacturing grew rapidly on the strength of import substitution.

This trend changed after the oil crisis of 1973. Domestic demand for manufactured goods decreased sharply in the aftermath of the crisis. The growth of the manufacturing sector abruptly slowed as Brazil entered a sustained economic recession. Manufacturing production fell nearly 15% between 1980 and 1983. Import substitution industrialization could no longer continue as a viable option. Brazilian trade policies had to be more liberal because access to external markets and investments was necessary for Brazilian manufacturing to survive over the long run. By the end of the 1980s, the economy recovered and domestic demand for manufactured goods started growing again. By the early 1990s, trade liberalization gained momentum and plans to reduce import tariffs materialized.

Between 1990 and 2000, Brazilian manufacturing embraced globalization (Figure 7.11). The economic stability that ensued, the availability of abundant natural resources, a large inexpensive labor force, and relaxed import policies changed Brazil into an attractive destination for foreign investments. Foreign trade significantly increased since then and Brazilian markets continue to be opened to foreign products, increasing

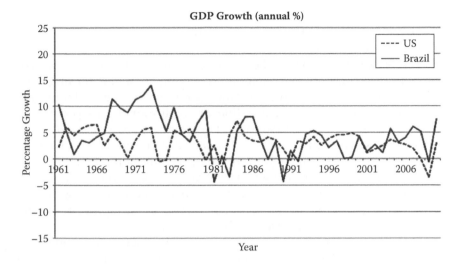

GDP Growth (annual %)

Figure 7.11 GDP growth comparison of Brazil and the United States. (*Source:* World Bank, 2012. World Development Indicators.)

competition to local manufacturing sectors. Brazilian manufacturing has undergone major institutional changes to adapt to global competition.

Brazilian manufacturing in the twenty-first century faces a number of challenges. First, a lack of quality in labor is a growing concern. The nation continues to change its educational structure and in-plant training policies to address this issue. Informal labor is an added impediment to industrial progress. The high costs of formal employment based on rigid labor laws and slow and inefficient documentation requirements continue to sustain the infamous trend to use informal labor in the manufacturing sector (Deloitte 2012).

The lack of process and product innovation is another crucial area that presents potential for improvement. Similar to all emerging economies, Brazilian manufacturing relies heavily on foreign technology for innovation. A shift in trend toward the manufacturing and export of high technology products is desirable for the future but is possible only with substantial investments in research and development activities. Brazilian exports continue to dominate in the areas of primary and labor-intensive low-tech goods. The rise in Chinese production of these items is a serious threat to Brazil.

Brazilian manufacturing also needs to improve its organization and structure in areas like logistics, communications, and transportation to compete with China. Lack of investment in developing local infrastructures has hurt the Brazilian manufacturing sector. However, Brazil's selection as the venue for major events like the 2014 FIFA World Cup and

2016 Summer Olympics will bring a huge influx of foreign cash and the required improvements to the infrastructure will provide a major boost to its manufacturing sector.

The Brazilian economy (like China's) is on the rise and as its people continue to improve their standard of living, demands for manufactured goods will continue to grow. However, Brazil is plagued by income inequalities. The unequal distribution of wealth among its population affects local demands. Internal markets continue to be polarized, weak and small. Externally, Brazil caters to the highly unpredictable markets of Europe and China. To circumvent this problem, Brazil must increase its market influence in growing and relatively stable markets in Asia and elsewhere in South America. Brazilian policies must be strengthened; high taxes and intermediary organization costs are crippling industrial progress and must be reformed to achieve real future progress.

On a positive note, Brazil has a well-established democracy and a government open to changing its policies to aid development of its manufacturing sector. Despite the need to improve labor quality, the Brazilian manufacturing sector has the advantage of a strong, culturally diverse and growing working population. Manufacturing activities have dominated in the areas of agro-based products and computer-related devices. On the strength of foreign investment, Brazil has established itself as a strong base for automobile and chemical-based industries. Sustainable energy will also play a major role in determining the competitiveness of its global manufacturing sector in the future. With sufficient natural resources Brazil has advantages for developing alternative renewable energy sources and biofuels. With the right set of changes, Brazil is capable of competing alongside China in the manufacturing sector.

7.3.2.2   Changes in Indian manufacturing sector

The Indian economy in the 1950s suffered heavily after its independence from British rule. Partition of the country into India and Pakistan was a major setback. The sudden change in geographical boundaries saw India lose a major share of its available natural resources to Pakistan. Manufacturing industries built completely on the strength of these raw materials were devastated. Major changes in administrative policies after independence and the rebuilding of a new economy under modern democratic rule showed promise.

The Indian manufacturing sector embraced import substitution policies and government-monitored industrialization was common in heavy industries like steel and mining. Despite the initial advantages of imitating foreign technology and promoting local manufacturing, import substitution continued to fail due to lack of competitiveness and poor productivity.

Between 1950 and 1970, manufacturing was an ailing industry because of the government's staunch support of socialistic policies (Rodrik and

Subramanian 2004). India's frequent involvement in wars and the government's failure to promote competitive capitalistic manufacturing policies led to huge economic deficits.

Indian businesses and industries suffered because of constant micro-level government intervention in areas such as labor and finance. The government failed to address the poor condition of its local infrastructure, especially transportation and communication. Bureaucratic inefficiencies and poor organization crippled India's manufacturing sector.

State planning failed to handle regular interruptions of manufacturing activities caused by natural disasters such as droughts, floods, and earthquakes that caused widespread power losses and damaged transportation and communications facilities. India needed a major overhaul of its industrial and economic policies. Its manufacturing sector showed no signs of recovery until the national economy was liberalized in 1990.

In the 1990s, as India suffered a major economic crisis, intense pressure from global organizations such as the International Monetary Fund led the government to overhaul its socialistic industrial policies in favor of privatization and capitalization. Between 1990 and 2000, India embraced globalization to a greater extent in the service sector than in manufacturing (Figure 7.12). While infrastructure continued to develop at a very slow rate, tax and bureaucracy reforms occurred faster as India fought to curb inflation. Import tariffs were reduced as Indian markets gradually opened to foreign products and—more importantly—to foreign technology.

Ailing government-controlled industries were privatized and opened to competition. Quantitative restrictions on domestic and foreign trade were removed and market competition was encouraged. The idea of developing special economic zones for manufacturing gained momentum.

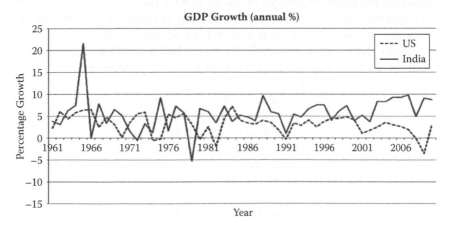

Figure 7.12 GDP growth comparison of India and the United States. (*Source:* World Bank 2012. World Development Indicators.)

Despite a temporary setback due to economic sanctions in 1998, India continued to grow on the strength of its IT industries and a steadily growing manufacturing sector.

One of the major changes on the part of the global manufacturing sector in the twenty-first century is the offshoring (or outsourcing) of manufacturing jobs from developed countries to countries with low labor costs. Indian manufacturing needs to take more advantage of the strength of its skillful, abundant, and inexpensive labor resources to capture a major share of offshoring opportunities. It currently lags behind China; Taiwan and Mexico are performing better than India in capturing offshored jobs.

India as an attractive destination for foreign direct investments is constrained by its limited infrastructure and rigid labor laws. Lack of constant power supplies and unpredictable supply chains due to poor transportation and communication continue to hurt the global image of Indian production. India must improve its local infrastructures if it wants to establish a strong manufacturing base for both local and offshored product lines.

India needs institutional reforms and must address sensitive issues like labor and capital policies headlong to make immediate and positive impacts on its manufacturing sector. The availability of institutional financing at low interest rates and government fiscal policies favoring manufacturing development are essential for future progress.

The thriving IT-based service sector in India has already markedly increased standards of living (Deloitte 2007). The growing middle class population is an indication of the huge potential of the Indian domestic market. The government must develop favorable market environments for its local manufacturing sector to exploit domestic demands for their goods. Diversity of the Indian market and the complexity of customer segmentation are other challenges of the manufacturing sector.

India lacks structural and organizational integration. Its manufacturing sector is littered with unregistered small- and medium-scale manufacturers that resist organization efforts. India needs to regulate and integrate these industries to provide wage and product price stability across its domestic market.

Indian manufacturing lacks product and process innovation. Despite its capacity to mass produce low-tech products, the sector should focus on quality and production efficiency for future success and also strive for operational excellence and improve resource management skills. Lack of intellectual property protection is another area of concern to be addressed if India is to attract foreign investments, especially in chemicals and pharmaceuticals.

India as an attractive destination for foreign manufacturing investments is growing on the strength of its abundant natural resources and raw materials and capacity to provide quality labor. The abundance of well-trained (and inexpensive) English-speaking scientists and engineers makes India a preferred candidate for foreign R&D investments in comparison to China and other Asian countries. If India is to emerge as an independent manufacturing powerhouse, its manufacturing sector must revamp its infrastructure and invest heavily in technology and R&D.

7.3.2.3   Changes in South African manufacturing sector

The success of South African manufacturing for over 50 years was built on the principle of protectionist policies like import substitution industrialization. Similar to emerging economies like Brazil, the biggest challenge for South African manufacturing in the 1940s was competing against superior products of established European and U.S. manufacturers. The most obvious way to curb economic imperialism and promote local industries was for the government to impose tight controls on its industrial tariff policies in favor of local manufacturing. Unlike other emerging economies, South Africa survived longer with its protectionist policies on the strength of its massive mineral resources (Scheider 2000). Much of South African industrialization resulted from the progress of its mining industries.

The first real sign of South African industrialization appeared in the 1920s. The government urged its mining industries to invest their massive profits in local manufacturing through a series of controlled tariff policies. The growth of manufacturing needed the support of primary industries like energy, iron, and steel production. Any private investment in manufacturing was unlikely without proper infrastructures and availability of primary goods. The South African government took the initiative and invested heavily in the primary sector to provide a proper platform for the manufacturing sector.

Between 1950 and 1960, the South African manufacturing sector grew under the protection of government tax policies. The government adopted a mixed trade policy under which import and export taxes were liberal in areas of export strength (like minerals) and tighter for consumer and intermediary manufactured goods. Government fiscal policies were framed with the intention of promoting manufacturing. Low-interest capital was available for new and expanding industries.

Another important factor behind the success of the South African mining and manufacturing sector in the 1950s and 1960s was its infamous exploitation of black labor. During that era, inexpensive, hard-working, and readily available labor contributed greatly to the growth of manufacturing industries.

During the 1970s, however, the situation began to change. South Africa and other countries were victims of "oil shock" as soaring energy prices took a toll on their manufacturing sectors. Local demands for manufactured goods disappeared because of the extremely low wages paid most blacks. The country's failure to develop indigenous technology, sudden decrease in domestic demand, failure to exploit new international

markets, political unrest, and global estrangement due to apartheid policies are only some of the reasons for the continued economic downturn
suffered by South African manufacturing.

From 1980 to 1990, manufacturing production and sales decreased
sharply and worker productivity reached an extreme low. Between 1990
and 2000, the end of apartheid brought much-needed political stability.
South African markets opened to foreign trade. The manufacturing sector
revived on the strength of recovering domestic demand and increases in
foreign reserves.

Throughout the twentieth century, South African manufacturing has
achieved mixed success (Figure 7.13). The nation continues to revamp its
industrial policies and manufacturing structure to suit global trade. Its strategic location linking the eastern and the western halves of the world and the
availability of strong manufacturing infrastructure have turned South Africa
into an attractive destination for direct foreign investments. Favorable bank-
ing systems and the availability of skilled labor working for lower wages
than their Europe and the U.S. counterparts are added advantages.

However, the growth of the manufacturing sector slowed by the start
of the twenty-first century due to saturation in domestic demand. South
Africa must actively stimulate exports as the key to expanding its manufacturing sector. The country continues to be a world leader in mineral wealth.

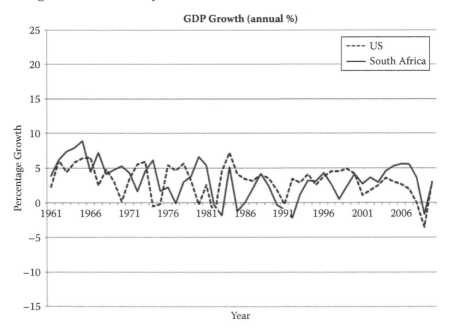

Figure 7.13 GDP growth comparison of South Africa and the United States.
(*Source:* World Bank 2012. World Development Indicators.)

The automotive industry is one of its largest and most successful economic engines. Other key areas in the manufacturing sector are metals, chemicals, information and communications, electronics, textiles, and footwear.

7.4   Productivity comparisons of various economic groups

7.4.1   Manufacturing output per hour indices

A number of graphs (Figures 7.14 through 7.18) compare the manufacturing output per hour indices of the United States, Japan, and Western European countries. These graphs follow a similar trend for reasons discussed in previous sections. For long periods before 1970, the manufacturing output per hour indices of Japan and Western European countries trailed behind indices of the United States. Between 1970 and 2000, Japanese and European manufacturing output per hour indices were slightly better than those for the United States. Since 2000, however, U.S. manufacturing output per hour gradually recovered (Brand South Africa 2012).

7.4.2   Total factor productivity

Total factor productivity (TFP) gives an overall measure of output based on unconventional non-production-based inputs such as innovation and technology. Figure 7.19 compares TFP ratings of BRICS, Japan, and Western Europe, and the United States. In the areas of innovation and technology,

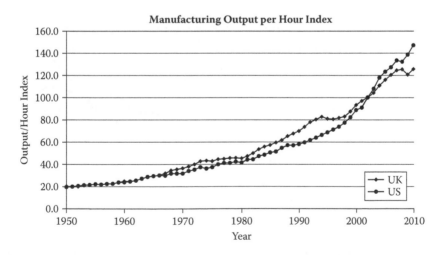

Figure 7.14 Comparison of U.K. and U.S. manufacturing output per hour index. (*Source:* U.S. Bureau of Labor Statistics 2011a. International Comparisons of Productivity and Unit Labor Cost Trends: 2010 Data Tables.)

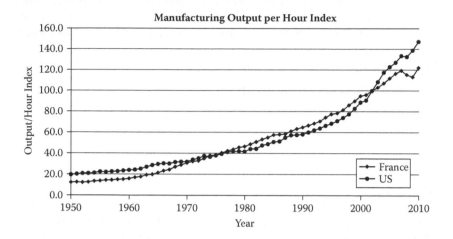

Figure 7.15 Comparison of France and U.S. manufacturing output per hour index. (*Source:* U.S. Bureau of Labor Statistics 2011a. International Comparisons of Productivity and Unit Labor Cost Trends: 2010 Data Tables.)

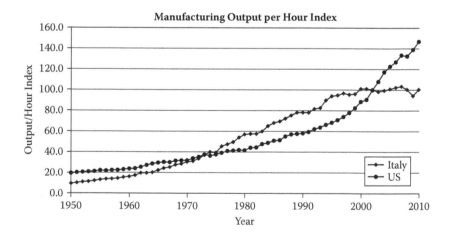

Figure 7.16 Comparison of Italy and U.S. manufacturing output per hour index. (*Source:* U.S. Bureau of Labor Statistics 2011a. International Comparisons of Productivity and Unit Labor Cost Trends: 2010 Data Tables.)

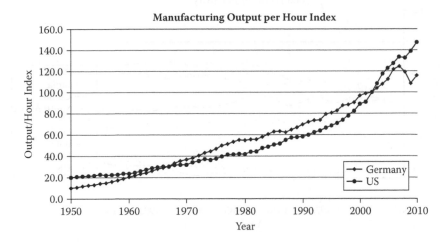

Figure 7.17 Comparison of Germany and U.S. manufacturing output per hour index. (*Source:* U.S. Bureau of Labor Statistics. 2011a. International Comparisons of Productivity and Unit Labor Cost Trends: 2010 Data Tables.)

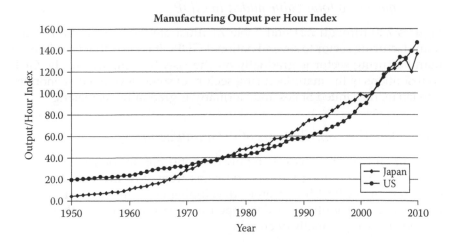

Figure 7.18 Comparison of Japan and U.S. manufacturing output per hour index. (*Source:* U.S. Bureau of Labor Statistics. 2011a. International Comparisons of Productivity and Unit Labor Cost Trends: 2010 Data Tables.)

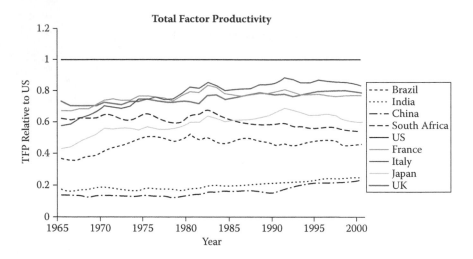

Figure 7.19 Total factor productivity levels with respect to the United States for 1965–2000. (*Source:* United Nations Statistics Division. 2012. http://unstats.un.org/unsd/default.htm accessed March 26, 2012.)

the United States and other developed economies still have upper hands over developing economies like China and India.

7.4.3   Percentage contribution of manufacturing output to total value added to GDP

Figures 7.20 through 7.24 and Table 7.1 illustrate the contribution of the manufacturing sector to several national GDPs. It is clear that the Chinese manufacturing sector is gradually on the rise. In comparison, the GDP contributions of the manufacturing sectors of some developed economies like Japan, the United States and Germany is gradually decreasing.

7.4.4   Relative contributions of the manufacturing sector

Table 7.2 and Figure 7.25 explain how the contributions of manufacturing sectors changed among various countries from 1970 through 2010. The contributions of BRICS and other developing economies are rising and the relative contributions of developed economies including that of the United States is gradually decreasing.

7.4.5   Labor productivity and manufacturing GDP

The developed economies dominate in the area of labor productivity (Figures 7.26 and 7.27). In terms of GDP per person employed, BRICS

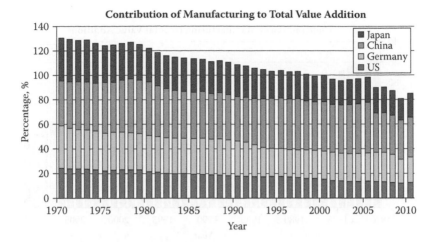

Figure 7.20 Percentage contribution of manufacturing to total value added to GDP for 1970–2010. Comparison of Japan, China, Germany, and the United States. (*Source:* United Nations Statistics Division. 2012. http://unstats.un.org/unsd/ default.htm accessed March 26, 2012.)

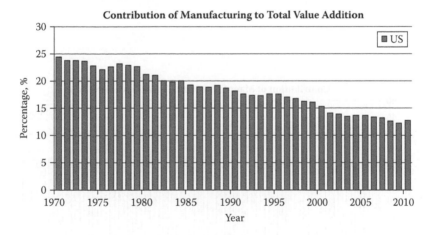

Figure 7.21 Percentage contribution of U.S. manufacturing sector to total value added to its GDP for 1970–2010. (*Source:* United Nations Statistics Division. 2012. http://unstats.un.org/unsd/default.htm accessed March 26, 2012.)

lag far behind developed economies like the United States, Japan, and Europe. This clearly indicates the technological advancements that augment labor productivity in the manufacturing sectors of these countries. However, it is wise to interpret labor productivity in conjunction with the percentage of workers employed in manufacturing. Offshoring manufacturing jobs is likely to increase the labor productivity of developed

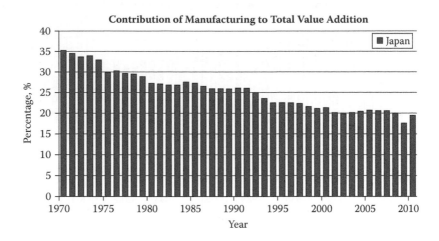

Figure 7.22 Percentage contribution of Japanese manufacturing sector to total value added to its GDP for 1970–2010. (*Source:* United Nations Statistics Division. 2012. http://unstats.un.org/unsd/default.htm accessed March 26, 2012.)

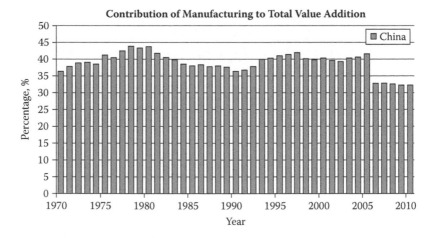

Figure 7.23a Percentage contributions of manufacturing sectors of BRICS countries to total values added to their respective GDPs for 1970–2010. (*Source:* United Nations Statistics Division. http://unstats.un.org/unsd/default.htm accessed March 26, 2012.) (continued)

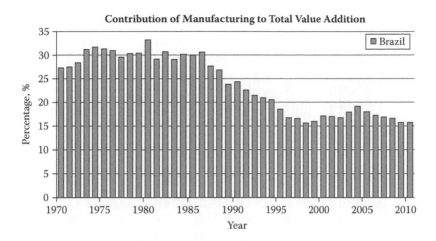

Figure 7.23b (continued) Percentage contributions of manufacturing sectors of BRICS countries to total values added to their respective GDPs for 1970–2010. (*Source:* United Nations Statistics Division. http://unstats.un.org/unsd/default. htm accessed March 26, 2012.) (continued)

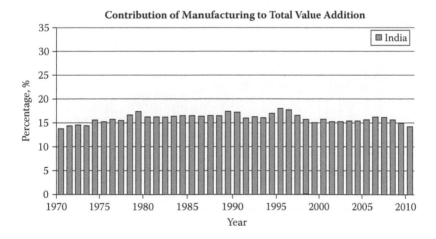

Figure 7.23c (continued) Percentage contributions of manufacturing sectors of BRICS countries to total values added to their respective GDPs for 1970–2010. (*Source:* United Nations Statistics Division. http://unstats.un.org/unsd/default. htm accessed March 26, 2012.) (continued)

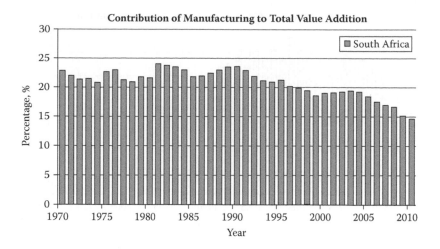

Figure 7.23d (continued) Percentage contributions of manufacturing sectors of BRICS countries to total values added to their respective GDPs for 1970–2010. (*Source:* United Nations Statistics Division. http://unstats.un.org/unsd/default. htm accessed March 26, 2012.)

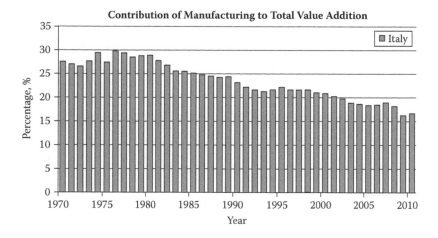

Figure 7.24a Percentage contribution of manufacturing sectors of Western European countries to total values added to their respective GDPs for 1970–2010. (*Source:* United Nations Statistics Division. 2012. http://unstats.un.org/unsd/default.htm accessed March 26, 2012.) (continued)

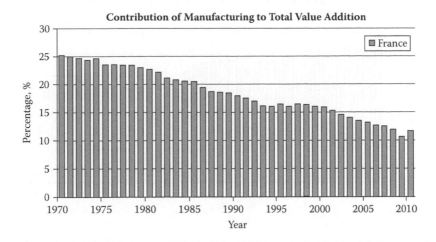

Figure 7.24b (continued) Percentage contribution of manufacturing sectors of Western European countries to total values added to their respective GDPs for 1970–2010. (*Source:* United Nations Statistics Division. 2012. http://unstats.un.org/ unsd/default.htm accessed March 26, 2012.) (continued)

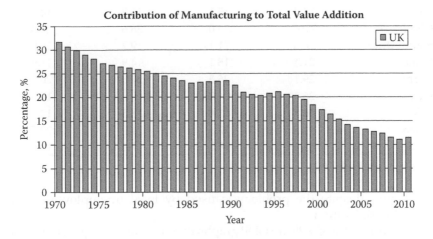

Figure 7.24c (continued) Percentage contribution of manufacturing sectors of Western European countries to total values added to their respective GDPs for 1970–2010. (*Source:* United Nations Statistics Division. 2012. http://unstats.un.org/ unsd/default.htm accessed March 26, 2012.) (continued)

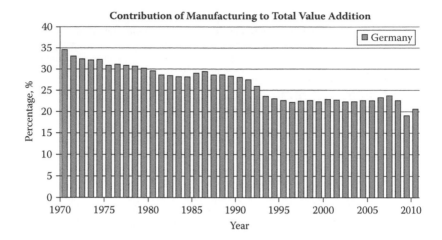

Figure 7.24d (continued) Percentage contribution of manufacturing sectors of Western European countries to total values added to their respective GDPs for 1970–2010. (*Source:* United Nations Statistics Division. 2012. http://unstats.un.org/unsd/default.htm accessed March 26, 2012.)

Table 7.1 Average Percentage Contributions of Manufacturing to Country GDPs

Country	1970–1979	1980–1989	1990–1999	2000–2010
Brazil	28.0	29.7	16.1	14.8
Russia	NA	NA	21.1	15.4
India	14.3	15.1	15.3	14.5
China	40.6	39.0	38.6	36.8
South Africa	20.6	21.1	19.2	16.0
France	21.4	18.1	14.9	11.9
Germany	29.1	26.3	21.8	20.2
Italy	26.3	24.0	19.8	16.9
U.K.	26.1	21.6	18.3	12.1
Japan	31.5	27.5	24.0	20.7
U.S.	23.0	19.7	17.1	13.6

Source: World Bank 2012.

economies despite the problems caused by local unemployment and de-industrialization.

Figures 7.28 and 7.29 are GDP graphs that illustrate the growing prominence of Chinese manufacturing. While the United States dominated the manufacturing sector for the past 30 years, China achieved a steep climb to overtake the U.S. manufacturing GDP. As of 2010, China's GDP was the largest in the world. The figures also indicate how the manufacturing GDPs of other BRICS countries gradually matched pace with Europe.

Table 7.2 Relative Percentage Contributions of Manufacturing Sectors, 1970–2010

Country	1970	1980	1990	2000	2010
Brazil	1.6	3.6	2.3	2.2	4.0
Russia	NA	NA	4.0	1.2	3.0
India	1.5	1.6	1.5	1.6	3.2
China	6.3	7.5	4.3	11.3	27.3
South Africa	0.7	0.9	0.7	0.5	0.7
France	6.1	7.9	5.9	4.4	3.8
Germany	12.3	14.1	13.0	9.2	8.7
Italy	5.1	7.1	7.1	4.8	4.4
U.K.	6.7	7.2	6.1	5.3	3.3
Japan	13.3	17.0	24.0	24.1	15.4
U.S.	46.5	33.1	30.9	35.4	26.3

Source: World Bank 2012.

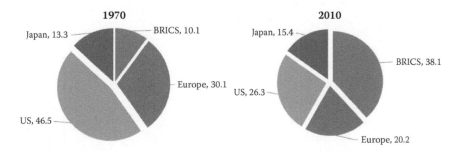

Figure 7.25 Relative percentage contributions of manufacturing sectors for 1970–2010. (*Source*: World Bank. 2012. World Development Indicators. http://publications.worldbank.org/index.php?main_page=page&id=8 accessed March 23, 2012.)

7.4.6   Other indicators

The World Bank and other national and international organizations track various indicators of economic progress. Figure 7.30 is a comparison of hourly wages compiled by the U.S. Bureau of Labor Statistics (2011b). Figures 7.31 through 7.33 respectively illustrate production growth rates, labor force utilization, and power consumption data for various countries. Tables 7.3 through 7.5 compare export volumes, export values, and changes over time for various countries.

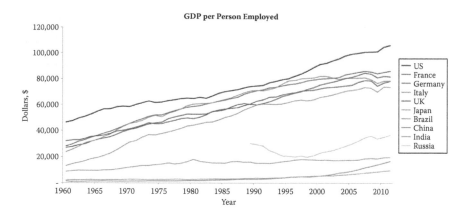

Figure 7.26 Relative percentage contributions of manufacturing sectors for 1960–2010. Data for Russian Federation available only after 1990s. (*Source:* World Bank. 2012. World Development Indicators. http://publications.worldbank.org/index. php?main_page=page&id=8 accessed March 23, 2012.)

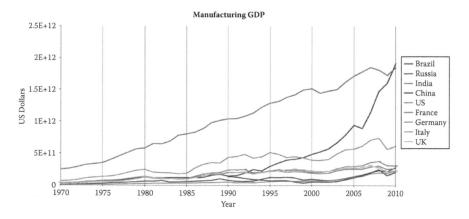

Figure 7.27 Relative percentage contributions of manufacturing sectors for 1970–2010. (*Source:* World Bank. 2012. World Development Indicators. http://publications.worldbank.org/index.php?main_page=page&id=8 accessed March 23, 2012.)

7.5 Conclusions

In summary, a study of various countries' economies led to several observations. First, technology and innovation continue to be the distinguishing assets of developed economies that reinvent their manufacturing strategies on the strength of advanced technology to dominate new markets with high-end products. Competition at the lower end of technology has significantly diminished or disappeared in developed countries. Developing economies are emerging as clear favorites. However, if they want to

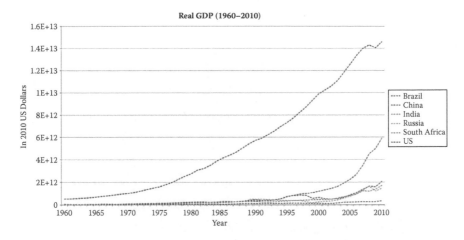

Figure 7.28 International comparison of real GDPs of the United States and BRICS for 1960–2010. (*Source:* World Bank. 2012. World Development Indicators. http://publications.worldbank.org/index.php?main_page=page&id=8   accessed March 23, 2012.)

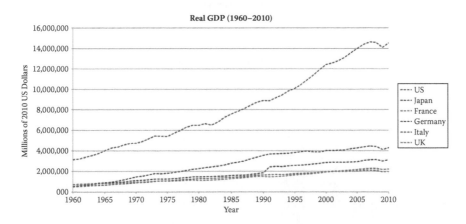

Figure 7.29 International comparison of real GDPs of the United States, Japan, and Europe for 1960–2010. (*Source:* World Bank. 2012. World Development Indicators. http://publications.worldbank.org/index.php?main_page=page&id=8   accessed March 23, 2012.)

increase their shares of benefits from the value chain, they must invest in research and development activities and bridge the technological gap.

Developing economies have an overwhelming advantage in the area of product cost because they have huge workforces willing to work for lower wages. Factory automation and robotics as measures to counter high labor

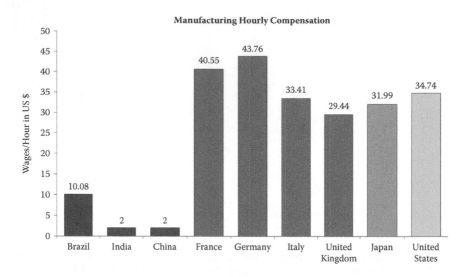

Figure 7.30 International comparison of hourly wages in manufacturing sector in 2010. (*Source:* U.S. Bureau of Labor Statistics. 2011a. International Comparisons of Hourly Compensation Costs in Manufacturing, 2010.)

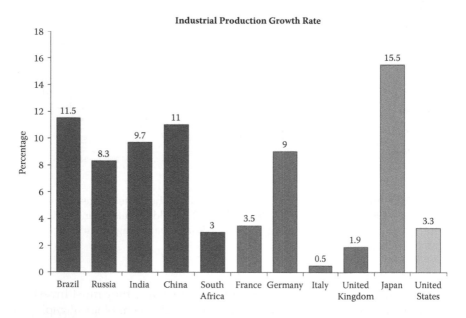

Figure 7.31 International comparison of production growth rate for 2011. (*Source:* U.S. Central Intelligence Agency. 2010. *World Fact Book*. https://www.cia.gov/library/publications/the-world-factbook/index.html accessed March 24, 2012.)

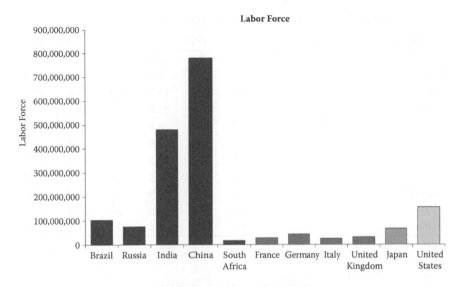

Figure 7.32 International comparison of labor force for 2011. (*Source:* U.S. Central Intelligence Agency. 2010. *World Fact Book.* https://www.cia.gov/library/publications/the-world-factbook/index.html accessed March 24, 2012.)

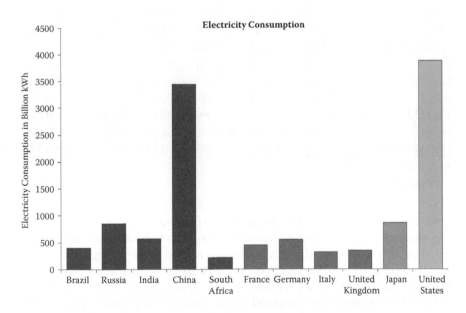

Figure 7.33 International comparison of electric power consumption for 2011. (*Source:* U.S. Central Intelligence Agency. 2010. *World Fact Book.* https://www.cia.gov/library/publications/the-world-factbook/index.html accessed March 24, 2012.)

Table 7.3 Percentage Changes in Volumes of Export Goods

Country	1980	1985	1990	1995	2000	2005	2010	2011
Brazil	25.7	2.2	–7.7	–2.0	11.1	9.4	9.5	2.9
Russia	n/a	n/a	n/a	7.7	7.9	4.7	8.5	4.7
India	1.4	0.6	9.7	12.2	13.5	10.8	19.8	12.4
China	21.4	6.5	–4.0	12.3	22.8	24.5	28.4	9.4
South Africa	0.6	10.8	–0.7	10.5	8.9	8.8	6.1	5.6
France	3.3	2.6	5.5	9.2	12.8	3.1	11.3	4.3
Germany	5.3	7.5	7.5	6.5	13.5	7.6	15.2	9.1
Italy	0.3	2.7	3.5	13.8	12.0	3.6	12.4	7.0
United Kingdom	1.3	5.5	6.2	9.8	11.5	8.6	10.8	5.1
Japan	8.8	5.2	5.0	3.3	8.8	0.7	25.4	–2.1
United States	11.9	3.7	8.4	11.7	11.1	7.5	14.4	7.4

Source: World Bank 2012.

Table 7.4 Values of Exports[a]

Country	1960	1970	1980	1990	2000	2011
Brazil	1,270	2,739	20,132	31,414	59,644	256,048
Russia	n/a	n/a	n/a	39,931	103,003	495,928
India	1,314	2,024	8,441	17,813	42,627	294,903
China	2,571	2,307	18,099	62,091	249,203	1,899,182
South Africa	1,985	3,344	25,540	23,568	30,430	93,097
France	6,854	18,098	116,031	216,466	323,528	584,449
Germany	11,382	34,233	192,865	409,286	548,962	1,391,939
Italy	3,636	13,184	77,682	170,495	236,668	515,414
United Kingdom	10,341	19,351	110,090	185,154	282,873	434,647
Japan	3,917	18,972	130,511	288,000	478,542	824,525
United States	20,540	43,239	221,073	393,106	772,280	1,480,727

Source: World Bank 2012.

[a] Millions of U.S. dollars.

costs have lost their attraction as developed economies continue to choose the alternative of offshoring manufacturing jobs to balance the books.

The threat of de-industrialization is real as labor composition continues to change in developed economies in favor of the service sector. Manufacturing jobs have been lost to developing countries at an alarming pace. It is necessary to understand the importance of protecting manufacturing jobs. Sustenance of the service industries depends on the manufacturing sector over the long term. This is visible in the growing GDP values and faster GDP growth rates of the manufacturing sectors of emerging economies.

Table 7.5 Percentage Changes in Values of Exports

Country	1960–1970	1970–1980	1980–1990	1990–2000	2000–2010
Brazil	53.64	86.39	35.91	47.33	76.71
Russia	n/a	n/a	n/a	61.23	79.23
India	35.11	76.02	52.61	58.21	85.55
China	–11.44	87.25	70.85	75.08	86.88
South Africa	40.65	86.91	–8.37	22.55	67.31
France	62.13	84.40	46.40	33.09	44.64
Germany	66.75	82.25	52.88	25.44	60.56
Italy	72.42	83.03	54.44	27.96	54.08
United Kingdom	46.56	82.42	40.54	34.55	34.92
Japan	79.35	85.46	54.68	39.82	41.96
United States	52.50	80.44	43.76	49.10	47.84

Source: World Bank 2012.

Developing economies lag in the areas of infrastructure improvements and organization and process innovation. They should look at the Japanese manufacturing techniques for improving production efficiency and operational excellence. If emerging economies want to reach global heights, they must strongly support and improve their manufacturing sectors.

Resource utilization and maximization are the fundamental drivers of modern industrialization. Both developed and developing countries have reaped a mixed share of the spoils. While developed countries managed to augment their resources with high-value additions based on the strength of their technologies, developing countries have failed somewhat in this area. However, despite long histories of industrialization, some developed countries (like Japan) are at the shallow ends of their natural resource pools and face a slight disadvantage in comparison to emerging economies.

Markets for manufactured goods in developed countries are relatively slow and already saturated. Conversely, developing countries boast vibrant markets hungry for manufactured goods. Globalization and liberal trade policies allow developed economies to continue to capture large shares of growing markets in emerging countries based on quality, but the risk of intellectual property theft remains a major impediment.

The United States still leads the world in manufacturing worker productivity. This lead is likely to continue for several reasons. As levels of technology and automation increase in BRICS, labor costs will also increase. This is already happening in China. BRICS will need far more skilled workers in the future to utilize new technologies effectively. They must invest in training and develop new technologies through investments in research and development. Perhaps the most crucial element is attempting to eliminate corruption and intellectual property abuse.

References

Banister, J. and Cook, G. 2011. China's employment and compensation costs in manufacturing through 2008. *Monthly Labor Review*, March, pp. 39–52.

Brand South Africa: Manufacturing in South Africa. Big Media Publishers. http://www.southafrica.info/business/economy/sectors/manufacturing.htm accessed March 21, 2012.

Crafts, N. and Mills, T.C. 2005. TFP growth in British and German manufacturing, 1950–1996. *Economic Journal*, 115, 649–670.

Deloitte. 2012. Competitive Brazil: Challenges and Strategies for the Manufacturing Industry. http://www.deloitte.com/assets/Dcom-Brazil/Local%20Assets/Documents/Ind%C3%BAstrias/Manufatura/livro_ingles.pdf. Accessed April 29, 2013.

Deloitte. 2011. Where is China's manufacturing industry going? China Manufacturing Competitiveness Study. http://www.deloitte.com/view/en_CN/cn/ind/mfg/5a71c737654d3310VgnVCM2000001b56f00aRCRD.htm. Accessed April 29, 2013.

Deloitte. 2010. Global Manufacturing Competitiveness Index. http://www.deloitte.com/view/en_GX/global/industries/manufacturing/a1a52c646d-069210VgnVCM200000bb42f00aRCRD.htm. Accessed April 29, 2013.

Deloitte. 2007. Globalizing Indian Manufacturing: Competing in Global Manufacturing and Service Networks. http://www.deloitte.com/view/en_GX/global/insights/deloitte-research/manufacturing-research/0b02eb f4c52fb110VgnVCM100000ba42f00aRCRD.htm. Accessed April 29, 2013.

Dower, H.J. 1990. Showa: The Japan of Hirohito. *Daedalus*, 119, 49–70.

European Commission. 2003. The Future of Manufacturing in Europe 2015–2020: The Challenge for Sustainability. Brussels: Joint Research Center.

European Commission. 2012. Eurostat: European Statistics. http://epp.eurostat.ec.europa.eu/portal/page/portal/statistics/themes accessed March 23, 2012.

Fagerberg, J., Guerrieri, P., and Verspagen, B. 1999. *The Economic Challenge for Europe: Adapting to Innovation-Based Growth*. London: Elgar.

Gordon, A. 2003. *A Modern History of Japan: From Tokugawa Times to the Present*. New York: Oxford University Press.

Hesse, D.M. and Tarkka, H. 1986. The demand for capital, labor and energy in European manufacturing industry before and after the oil price shocks. *Scandinavian Journal of Economics*, 88, 529–546.

Hitomi, K. 1993. Manufacturing technology in Japan. *Journal of Manufacturing Systems*, 12, 209–215.

Japan Ministry of Trade and Industry. 2010. Japan's Manufacturing Industry.

Johansson, U. 2008. The main features of the EU manufacturing industry: Eurostat statistics in focus. 37, 1–8. http://epp.eurostat.ec.europa.eu/cache/ITY_OFFPUB/KS-SF-08-037/EN/KS-SF-08-037-EN.PDF. Accessed April 29, 2013.

Johnson, C.A. 1982. *MITI and the Japanese Miracle: The Growth of Industrial Policy 1925–1975*. Palo Alto, CA: Stanford University Press.

Mosk, C. 2001. *Japanese Industrial History: Technology, Urbanization and Industrial Growth*. Armonk, NY, M.E. Sharp.

Ohno, K. 2000. Globalization *of Developing Countries: Is Autonomous Development Possible?* Tokyo: Toyo Keizai Shimposha.

Okada, Y. 1999. *Japan's Industrial Technology Development: The Role of Cooperative Learning and Institutions.* Heidelberg: Springer.

Rodrik, D. and Subramanian, A. 2004. From Hindu growth to productivity surge: The mystery of the Indian growth transition. NBER Working Paper 10376.

Scheider, G.E. 2000. The development of the manufacturing sector in South Africa. *Journal of Economic Issues.* 34(2), 413–424.

Shoten, I. 1989. *The Economic History of Japan, 1988–1989*, 8 volumes. Tokyo: Nihon Keizaishi.

United Nations Statistics Division. http://unstats.un.org/unsd/default.htm accessed March 26, 2012.

U.S. Bureau of Labor Statistics. 2011a. International Comparisons of Productivity and Unit Labor Cost Trends: 2010 Data Tables.

U.S. Bureau of Labor Statistics. 2011b. International Comparisons of Hourly Compensation Costs in Manufacturing, 2010.

U.S. Central Intelligence Agency. 2010. *World Fact Book.* https://www.cia.gov/library/publications/the-world-factbook/index.html accessed March 24, 2012.

Ward, J.R. 1994. The industrial revolution and British imperialism, 1750–1850. *Economic History Review*, 47, 44–65.

World Bank, 2012. World Development Indicators. http://publications.worldbank.org/index.php?main_page=page&id=8 accessed March 23, 2012.

Zheng, Y. 2005. *Productivity Performance in Developing Countries: Country Case Studies.* Beijing: People's Republic of China.

chapter eight

Benefits of advanced manufacturing technologies

Jorge Luis García Alcaraz, Sergio Gutiérrez Martínez, and Salvador Noriega Morales

Contents

8.1   Introduction

A general definition of advanced manufacturing technology (AMT) was given by Zairi (1992). He describes AMT as a social–technical system that requires continued revisions, readjustments, and changes to adapt to the requirements of the competitive world (flexibility). This definition is very general, and lacks precision. The glossary of statistic terms of the Organisation for Economic Cooperation and Development (OECD 2011) defines it as equipment controlled by a computer or microelectronics and applied to design, manufacturing, or product manipulation.

The literature contains several descriptions of AMT and uses terms such as "integrated manufacturing systems" to describe the use of computers, but not all manufacturing technologies include computing systems. Therefore, misconceptions prevail concerning the distinction between the two types of technologies designated hard and soft. Hard technology incorporates computing systems in contrast to soft technology in which the control of operations generally involves methods utilizing administrative tools. The result is a lack of agreement about the most fundamental aspect of AMT. The definition continues to be a source of confusion. The diversity of benefits and the complexity of the components of AMT contribute further to this confusion.

Among several taxonomies proposed for AMT, the most accepted classification was proposed by Small and Yasin (1997). They divided AMT into hard and soft technologies and subdivided the hard technologies into the following components of a production line:

- Robots
- Computer-aided design (CAD)
- Computer-aided manufacturing (CAM)
- Computer-aided engineering (CAE)
- Computer-integrated manufacturing (CIM)
- Computer numerical control (CNC)
- Flexible manufacturing systems (FMSs)
- Three-dimensional (3D) digitalization
- Fast prototypes
- Local area networks (LANs)
- Wide area networks (WANs)
- Technology information and communication (TIC)
- Industrial automation
- Automated guided vehicles (AGVs)
- Automated inspections (AIs)
- Artificial intelligence
- Laser technologies
- Electronic data interchange (EDI)
- Computer-aided process planning (CAPP)
- Automatic item loads and download
- Automated tool changes
- Computer-aided inspection, testing, and tracking
- Automated item identification (bar coding)
- Supervisory control and data acquisition (SCADA)

Soft technologies are techniques or methodologies applied to a production system to increase its effectiveness. Sometimes soft technologies are simply concepts or philosophies, examples of which are given below:

- Just-in-Time (JIT) production
- Manufacturing Resource Planning (MRP II)
- Enterprise Resource Planning (ERP)

- Group Technology (GT)
- Manufacturing Cells (MCs)
- Total Quality Management (TQM)
- Statistic Quality and Process Control (SQ/PC)
- Single Minute Exchange of Die (SMED).
- Total Productive Maintenance (TPM)
- Manufacturing Technique: Lean Manufacturing.

When these technologies are applied to production systems, several benefits are reported but their levels vary with the reporting authors. Table 8.1 lists the benefits mentioned by several authors. The benefits of management control and improvement in work environment are cited most frequently (in seven of eight references).

The benefits related to the flexibility of a production system are also described. AMT allows a greater family of products to be produced. The same applies to the response capability of engineering changes that improve productivity through design modifications. Another benefit is improvement of a company's image. All these benefits were reported in six of eight references.

The less reported benefits are stability of operations, enhanced capability to apply engineering changes (different from responding to engineering changes), and billing efficiency. A cursory review of Table 8.1 indicates a lack of agreement among experts regarding the benefits, although no contradictions were noted.

This research analyzed benefits achieved in industrial plants in Ciudad Juarez, Chihuahua, Mexico—one of the largest industrial areas in the country—and considers that the benefits of AMT projects may not have been documented adequately. It is important to mention that the Juarez sector represents 50% of Mexican exports of automotive manufacturing products. Also, the Aśociation de Maquiladoras A.C. (AMAC) represents 352 enterprise members in different areas. INEGI data (2010) indicate the economic importance of the Ciudad Juarez region.

The objective of this project was to identify the benefits that industrial plants in Juarez expected from AMT planning and those that were really obtained after AMT was installed and utilized. The results represent empirical evidence.

8.2   Methodology

8.2.1   First phase: Development of questionnaire

The initial phase was the review of literature to identify and list the papers related to the benefits of AMT. The main benefits reported are presented in Table 8.1. The benefits were used to build a questionnaire containing 32

Table 8.1 Benefits of AMT Investments Cited in Literature

Benefit Attainable with AMT	A	B	C	D	E	F	G	H	Total Citations	**Total**
Improvement in management control	*	*	*		*	*	*	*	7	
Improvement in work environment	*	*	*	*	*	*		*	7	
Flexibility	*		*	*	*	*	*		6	
Expansion in product line and depth		*	*		*	*	*	*	6	
Improved capability to respond to engineering changes	*	*		*	*		*	*	6	
Improved company image		*	*	*	*		*	*	6	
Improved response to variability of products mix	*	*			*	*	*		5	
Better integration of technology through function		*	*		*	*	*		5	
Improved attitude of workforce	*				*	*	*	*	5	
Improved capability to respond to supplier quality variabilities	*	*	*	*	*				5	
Improvement of management attitude	*		*		*	*		*	5	
Reduction of product development time		*	*		*	*	*		5	
Better work relations		*		*	*	*		*	5	

(continued)

Table 8.1 Benefits of AMT Investments Cited in Literature (continued)

Benefit Attainable with AMT	Author								Total Citations	**Total**
	A	B	C	D	E	F	G	H		
Improved capability to respond to delivery time variabilities of suppliers				*	*	*	*	*	5	
Overcoming deficiencies caused by lack of production management ability	*	*	*		*	*			5	
Improved responses to product changes		*	*		*	*			4	
Reduction of WIP inventories	*				*	*		*	4	
Better manufacturing process integration			*	*	*		*		4	
Outperforming ability deficiencies		*		*		*		*		4
Lower forecasting costs	*			*		*		*	4	
Enhanced operation stability	*		*					*	3	
Larger capacity to apply engineering changes	*				*				2	
Faster invoicing	*					*			2	

Author Key:

A. Kakati 1997.

B. Beaumont and Schroder 1997.

C. Noori 1997.

D. Swink and Nair 1997.

E. Millen and Sohal 1998.

F. Stock and McDermott 2001

G. Efstathiades et al. 2002.

H. Dangayach and Deshmukh 2006.

items related to design, engineering and technical, marketing, and process flexibility issues.

The initial survey was administered to 26 engineers working in manufacturing industries. This test survey led to several modifications of the questions before validation. A blank space was left after each question to allow the subjects to report additional benefits or observation. Finally, 10 benefits were eliminated from the initial questionnaire and 13 new ones added, resulting in a final survey of 35 items divided into categories of design, technical, marketing, and cost. The questionnaire had two columns, one for "before" and the other for "after" deployment of the AMT. A five-point Likert scale was used to assess the answers. A 1 indicated a total lack of benefit and a 5 indicated that the AMT benefit was acquired successfully (Likert, 1932).

8.2.2   Second phase: Survey administration

During the second phase, a total of 241 engineers working in the industrial sector in engineering, management, and supervisory positions were contacted and interviewed. The list of engineers came from the Aśociation de Maquiladoras, A.C. (AMAC) of Ciudad Juarez, Chihuahua, Mexico. Appointments were made with the engineers and the questionnaires were completed in their workplaces. Three visits per subject were considered suitable for obtaining satisfactory answers. After three visits or three email reminders without responses, some subjects were eliminated from the study.

8.2.3   Third phase: Analysis

Two databases were developed and analyzed using the Statistical Package for Social Sciences (SPSS Version 17). One related to the planned or expected benefits, and the other covered real benefits obtained from AMT. Cronbach´s alpha test was applied and values higher than 0.8 were used for validation. These items values were also analyzed about their impact on benefits; so they were eliminated from the questionnaire (Nunnally and Bernstein 2005, Cronbach 1951).

The validated questionnaire was accompanied by a descriptive analysis of the problems and the median was determined since the data were ordinal. The 25th and 75th percentiles were also calculated to find the interquartile range as a position measure and data dispersion (Denneberg and Grabisch 2004, Pollandt and Wille 2005, Tastle and Wierman 2007).

A correlation matrix was developed to determine the feasibility of a factorial analysis. Most correlations were higher than 0.3. The anti-image correlations were also analyzed. The Kaiser-Meyer-Olkin (KMO) index was calculated since it is recommended for indices exceeding 0.75. The Bartlett sphericity test was applied to measure the sample adequacy and the commonality was analyzed for every attribute (Nunnally 1978, Nunnally and Bernstein 2005).

A factorial analysis was conducted to determine the factors by means of the principal component analysis method using the correlation matrix to extract the components. The factors considered important had values equal to or greater than one or half of their eigenvalues. This search was limited to 100 iterations for the convergence of the result (Streiner and Norman 1995). For a better interpretation of critical factors, a factor rotation by the Varimax method (Lévy and Varela 2003) was done. The activities that integrated the factors were those with the highest factor load value, which is a correlation measure with the orthogonal factorial axis (Nunnally and Bernstein 2005).

8.3   Results

8.3.1   Questionnaire validation

The results of the analysis for questionnaire validation are shown in Table 8.2. The obtained values are higher than the 0.8 inferior limit, demonstrating that the questionnaire was adequate for obtaining the desired information.

8.3.2   Sample composition

A total of 60 surveys were received from different companies. The industry sectors are illustrated in Figure 8.1. The automotive sector was the most polled, with 22 surveys, followed by the electronics and plastic sectors with 11 and 9, respectively. The other replies were from building materials and medical sectors. Nine enterprises did not report their sectors.

The enterprises stated that robots integrated into their production systems were the most common recently acquired technologies (10 cases). Similarly, milling machines, vision systems, and generic equipment CNC, had 9, 8, and 7 answers respectively. The systems in which less investment was reported are soft ones such as the 5S systems, manufacturing cells, and visual aids. Figure 8.2 illustrates the details of recently acquired AMTs.

Table 8.2 Questionnaire Validation

Parameter	Planned Benefit		Obtained Benefit	
Cronbach's alpha	0.963		0.968	
Two halves test	First half	0.931	First half	0.937
	Second half	0.947	Second half	0.957

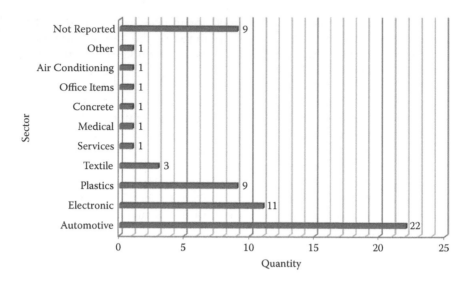

Figure 8.1 Sectors surveyed.

Respondents came from different areas or departments of their companies and all of them had participated in AMT evaluation and selection. Figure 8.3 illustrates the departments and respondent distribution. Engineering departments were very involved in determining AMT investments; 20 of the 60 respondents worked in engineering. Overall, it appears that these types of investment issues involve several departments and it follows the economic aspect is not the only one considered in the evaluation process. Only one auditor from an accounting department

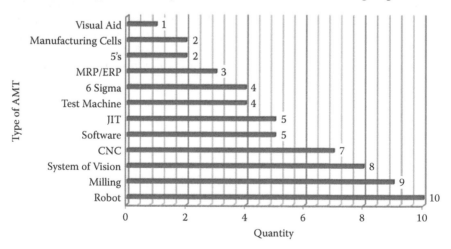

Figure 8.2 Results for recently acquired AMTs.

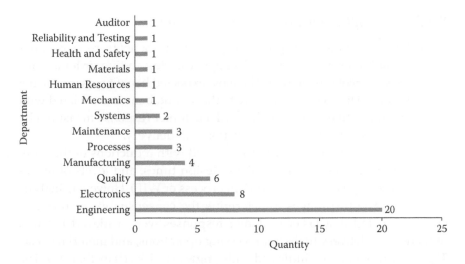

Figure 8.3 Departments involved in AMT investment decisions.

participated. However, it should be noted that most companies utilize different organizational structures based on their sizes. For example, we found that reliability and testing operations were integrated into quality departments in two medium companies.

An in-depth analysis of the distribution of the departments involved in AMT investment evaluations (e.g., grouping engineering, electronics, quality, manufacturing, maintenance, and production) reveals that most of the departments involved in making these decisions focused on purely technical aspects. Only one department noted participation of human resources in decision making. This indicates that the needs of workers were given little importance when the acquisition of new AMTs was considered.

In Mexico and many other countries, the integration of women into productive work sectors has gained great support. Table 8.3 shows the distribution of gender in relation to job tenure. Eleven enterprises employed 50 to 250 workers; 46 enterprises were larger and employed more than 250 workers; 3 enterprises did not report their sizes.

Table 8.3 Distribution of Gender and Work Tenures

Gender	Years in Job				Total
	1	2	5	10	
Male	8	19	9	9	45
Female	4	3	4	1	12
Total	12	22	13	10	57

8.3.3   Descriptive analysis: Expected benefits

Table 8.4 presents a descriptive analysis of the benefits realized after implementation of AMT. The results appear in descending order according to the values of the median. The most expected benefit was increasing market share., This is related closely to the second benefit associated with levels of competitiveness based on quick customer response measured by reduction of the time between production and delivery.

Fewer expected benefits were associated with reductions of the numbers of machines on production lines, design times, and levels of inventory in process (also known as work in process or WIP). However, analysis of interquartile (IR) ranges to determine the consensus in relation to a benefit revealed that faster customer responses were achieved through reductions in delivery times, reprocessing operations, and material waste. These low IR values are indicated with single asterisks (*) in Table 8.4. The highest IR values were achieved where no generalized consensus on the value of a benefit was indicated (reduction of production lot size, better engineering experience, issues for maintaining competitiveness). Double asterisks (**) indicate these high values in Table 8.4.

8.3.4   Descriptive analysis: Obtained benefits

Table 8.5 presents a descriptive analysis of the benefits of AMT as measured by their median values in descending order. The most commonly reported benefits relate to reprocessing reductions and deal with waste. AMT helps sustain market competitiveness. Additionally, a better engineering experience is achieved and product quality improves based on better reliability. Lesser benefits concerned reductions of workforce cost, machinery, WIP, and inventory rotation. The high costs may be due mainly to the use and operation of AMT that requires more skilled and specialized personnel earning higher salaries.

The attributes with lower IR values (indicating consensus in median results) such as reductions of WIP and time between conceptualization and manufacture, sustained competitiveness, less reprocessing, minimizing waste, and achieving quality improvement also show that the first four attributes with lower IRs have the largest medians. Therefore, it can be concluded that these attributes are important. In Table 8.5, the same asterisk marking system from Table 8.4 is used. The items that displayed high IR values (low consensus and an unclear value) were reductions in machinery and production lot size, better use of work space, and reductions in parts varieties.

Table 8.4 Descriptive Analysis: Expected Benefits

Benefit	Median	Percentiles 25	75	IR
Larger market coverage	3.308	2.517	3.974	1.457
Maintain level of competitiveness	3.273	2.389	4.068	1.679**
Time reduction between order reception and delivery time	3.27	2.429	3.973	1.544
Higher product quality	3.256	2.438	3.974	1.537
Higher plant capability	3.231	2.339	3.936	1.597
Fast responses to customer needs	3.22	2.469	3.854	1.385*
Higher feasibility	3.205	2.412	3.923	1.511
Higher sales	3.194	2.308	3.968	1.66
Better managerial experience	3.179	2.394	3.872	1.478
Reduction of delivery time	3.15	2.412	3.8	1.388*
Early introduction to market	3.118	2.344	3.853	1.509
Reduction of adjustments time (system and AMT)	3.105	2.329	3.829	1.5
Reduction of reprocessing time and scrap	3.098	2.347	3.768	1.421*
Improved plant utilization	3.079	2.265	3.816	1.551
Better engineering administration	3	2.172	3.855	1.683**
Reduction of production lot size	3	1.975	3.809	1.834**
Machinery flexibility	2.974	2.25	3.697	1.447
Reduction of process time	2.973	2.216	3.711	1.494
Enhancement of flexibility	2.973	2.23	3.697	1.468
Better organization of production operations	2.972	2.194	3.844	1.649
Higher labor productivity	2.971	2.214	3.708	1.494
Process flexibility	2.946	2.203	3.708	1.506
Volume flexibility	2.943	2.157	3.729	1.571
Improvement of inventory rotation	2.939	2.136	3.721	1.584
Design quality	2.893	2.054	3.732	1.679**
Reduction of manual material handling	2.889	2.139	3.657	1.518
Reduction of variety of products and parts	2.886	2.129	3.662	1.533
Reduction of production cost	2.868	2.145	3.625	1.48
Reduction of part variety	2.838	2.108	3.618	1.51
Reduction of labor cost	2.829	2.057	3.724	1.667
Effective use of floor space	2.825	2.125	3.636	1.511
Reduction of time from concept to production	2.824	2.132	3.603	1.471
Reduction of number of machines	2.811	2.068	3.603	1.535
Reduction of design time lapse	2.806	2.048	3.7	1.652
Reduction of inventory in process	2.78	2.11	3.561	1.451

* Low IR values. ** High IR values.

Table 8.5 Descriptive Analysis: Obtained Benefits

		Quartile		
	Median	25	75	IR
Reprocessing and waste reduction	4.348	3.589	4.946	1.356*
Maintain competitiveness level	4.326	3.558	4.919	1.361*
Better engineering experience	4.317	3.458	4.951	1.493
Increase of product quality	4.298	3.534	4.883	1.348*
Reliability increases	4.244	3.468	4.856	1.388
Volume flexibility	4.186	3.304	4.826	1.522
Reduction of time from conceptualization to manufacture of product	4.179	3.411	4.782	1.371*
Faster responses to consumer needs	4.15	3.371	4.788	1.417
Time reduction between order and delivery time	4.15	3.286	4.8	1.514
Increase of plant capability	4.125	3.167	4.813	1.646
Better design quality	4.108	3.31	4.757	1.446
Better organization for production operations	4.077	3.155	4.782	1.627
Process flexibility enhancement	4.071	3.279	4.726	1.447
Reduction of design time	4.056	3.29	4.722	1.432
Increased market share	4.051	3.25	4.718	1.468
Increase of overall flexibility	4.051	3.177	4.756	1.579
Increase of labor productivity	4.05	3.25	4.7	1.45
Reduction of processing time	4.049	3.227	4.72	1.492
Improved plant utilization	4.026	3.069	4.744	1.675
Better administration experience	4.025	3.227	4.688	1.46
Sales increases	4	3.04	4.727	1.687
Reduction of adjustment time (system and AMT)	3.969	3.125	4.684	1.559
Reduction of delivery time	3.941	3.176	4.649	1.472
Reduction of material handling	3.914	3.143	4.615	1.473
Reduction of production cost	3.868	3.158	4.579	1.421
Faster and earlier entrance to market	3.867	3.033	4.636	1.603
Better use of space	3.862	2.842	4.686	1.844**
Reduction in varieties of parts and products	3.844	3.016	4.614	1.599
Reduction in varieties of parts	3.833	2.818	4.611	1.793**
Machinery flexibility	3.824	3.015	4.632	1.618
Reduction of production lot size	3.793	2.765	4.656	1.892**
Improved inventory rotation	3.788	2.969	4.591	1.622
Reduction of WIP inventories	3.78	3.11	4.487	1.377*
Reduction in numbers of machines	3.759	2.667	4.606	1.939**
Reduction of cost of workforce	3.649	2.9	4.433	1.533

* Low IR values. ** High IR values.

8.3.5   Factorial analysis: Expected benefits

The feasibility analysis of the factorial analysis yielded a Kaiser-Meyer-Olkin (KMO) value of 0.777, indicating acceptable sample adequacy. Bartlett's sphericity test gave a chi-square value of 1081.37 with 595 degrees of freedom, indicating a significance of 0.00. This allowed us to conclude that factorial analysis could be applied appropriately. The results indicate that only seven factors explained 74.43% of the total variance. Tables 8.6 and 8.7 illustrate the results of the factorial analysis.

8.4   Conclusions

After the analysis of surveys from 60 enterprises in various industrial sectors of Ciudad Juarez, we concluded that the enterprises that invested in AMT achieved better market shares. These results agree with the findings of other researchers.

AMT improves soft production technologies and technical aspects of production operations, improves marketing measures, reduces time and material wastes during processing, improves quality, and enhances the effectiveness of knowledge management, increases the design efficiency and use of work space, and improves inventory handling.

Enterprises that implemented AMT achieved reductions of reprocessing operations and waste, were aided in maintaining their competitiveness levels, achieved better engineering experiences, improved product quality and reliability, and finally improved their flexibility to better respond to production volume changes. They also achieved improvements in the techniques and operations aspects of their production systems, increments in operative flexibility, marketing benefits, reductions of production cost and time, improvements in design of parts and products, and finally significant improvements of overall quality.

Table 8.6 Factorial Analysis for Expected Benefits

Benefit	FC	Factor
Reduction in numbers of machines	0.816	
Reduction in production lot size	0.781	
Reduction of the varieties of products and parts	0.771	Techniques and operative aspects 16.16%
Reduction in the varieties of parts	0.706	
Improvement in inventory rotation	0.694	
Market share increase	0.819	
Faster responses to consumer needs	0.785	
Reduction of delivery times	0.772	
Reduction of time between order receipt and delivery	0.772	Marketing aspects 16.01%
Increase of sales	0.753	
Earlier entrance to market	0.745	
Maintain competitiveness level	0.613	
Reduction of adjustment time (system and AMT)	0.793	
Reduction of reprocessing and waste	0.65	Waste reduction (time and material) 16.16%
Better organization of production	0.648	
Better engineering experience	0.749	
Better administration experience	0.671	
Increase of labor productivity	0.643	Quality and knowledge management 10.80%
Enhancement of reliability	0.626	
Increase of product quality	0.608	
Quality in design	0.864	
Reduction of design time	0.808	
Reduction of time between conceptualization and product manufacture	0.688	Design aspects 9.27%
Improved plant utilization	0.708	
Increase of plant capability	0.688	Space usage 7.36%
Reduction of WIP	0.627	Inventory 3.35%

Table 8.7 Factorial Analysis for Obtained Benefits

Benefit	FC	Factor
Reduction of variety of parts	0.787	
Reduction of variety of products and parts	0.728	
Improved inventory rotation	0.704	Technical and operative aspects 17.99%
Reduction of material handling	0.693	
Reduction of production lot size	0.651	
Reduction of numbers of machines	0.631	
Process flexibility	0.776	
Machinery flexibility	0.729	
Reduction of reprocessing and waste	0.721	Flexibility 15.64%
Reduction of WIP	0.697	
Volume flexibility	0.679	
Fast responses to consumer needs	0.837	
Reduction of delivery times	0.729	Marketing aspects 12.08%
Time reduction between order receipt and delivery	0.726	
Reduction of workforce cost	0.799	
Reduction of processing time	0.754	Cost and time reduction 7.54%
Reduction of cost of production	0.726	
Reduction of design time	0.828	
Design quality	0.751	
Reduction of time between conceptualization and product manufacture	0.654	Design aspects 7.48%
Reliability increment	0.782	Quality 6.23%
Increment in product quality	0.759	

References

Beaumont, N.B. and Schroder, R.M. 1997. Technology, manufacturing performance and business performance amongst Australian manufacturers. *Technovation*, 17, 297–307.

Cronbach, L.J. 1951. Coefficient alpha and the internal structure of tests. *Psychometrika*, 16, 297–334.

Dangayach, G. and Deshmukh, S. 2006. An exploratory study of manufacturing strategy practices of machinery manufacturing companies in India. *Omega*, 34, 254–273.

Denneberg, D. and Grabisch, M. 2004. Measure and integral with purely ordinal scales. *Journal of Mathematical Psychology*, 48, 15–22.

Efstathiades, A., Tassou, S.. and Antoniou, A. 2002. Strategic planning, transfer and implementation of advanced manufacturing technologies (AMTs): development of an integrated process plan. *Technovation*, 22, 201–212.

INEGI. 2010. Mexico National Institute of Statistics and Geography. Consulted online at: http://www.inegi.org.mx accessed September 10, 2011.

Kakati, M. 1997. Strategic evaluation of advanced manufacturing technology. *International Journal of Production Economics*, 53, 141–156.

Lévy, J.P. and Varela, M. 2003. *Análisis Multivariable para las Ciencias Sociales*, Madrid: Prentice Hall.

Likert, R. 1932. A technique for the measurement of attitudes. *Archives of Psychology*, 22, 1–55.

Millen, R. and Sohal, A. 1998. Planning processes for advanced manufacturing technology by large American manufacturers. *Technovation*, 18, 741–750.

Noori, H. 1997. Implementing advanced manufacturing technology: the perspective of a newly industrialized country (Malaysia). *Journal of High Technology Management Research*, 8, 1–20.

Nunnally, J.C. 1978. *Psychometric Theory*. New York: McGraw-Hill.

Nunnally, J.C. and Bernstein, H. 2005. *Teoría Psicométrica*, México: McGraw-Hill Interamericana.

OECD. 2011. *Glossary of Statistical Terms*. Geneva: Organisation for Economic Cooperation and Development. http://stats.oecd.org/glossary/detail. asp?ID = 52 accessed June 12, 2011.

Pollandt, S. and Wille, R. 2005. Functorial scaling of ordinal data. *Discrete Applied Mathematics*, 147, 101–111.

Small, M.H. and Yasin, M.M. 1997. Advanced manufacturing technology: implementation policy and performance. *Journal of Operations Management*, 15, 349–370.

Stock, G.N. and McDermott, C.M. 2001. Organizational and strategic predictors of manufacturing technology implementation success: an exploratory study. *Technovation*, 21, 625–636.

Streiner, D. and Norman, G.R. 1995. *Health Measurement Scales: A Practical Guide to Their Development and Use*, 2nd ed. Oxford: Oxford University Press.

Swink, M. and Nair, A. 2007. Capturing the competitive advantages of AMT: design–manufacturing integration as a complementary asset. *Journal of Operations Management*, 25, 736–754.

Tastle, W.J. and Wierman, M.J. 2007. Using consensus to measure weighted targeted agreement. *Fuzzy Information Processing Society*, 24, 31–35.

Zairi, M. 1992. Measuring success in AMT implementation using customer–supplier interaction criteria. *International Journal of Operations and Production Management*, 12, 34–55.

chapter nine

Productivity and technology
Techniques for industrial energy savings

Alina Adriana Minea

Contents

9.1   Introduction

9.1.1   *Important questions dealing with productivity issues*

When productivity concepts are discussed, five important questions arise. Before starting to talk about productivity, let us find some answers to these five issues.

What is productivity? Technically, productivity is the ratio of output to input. It is a measure of how efficiently and effectively a business or economy uses inputs such as labor and capital to produce outputs such as goods and services. An increase in productivity means that more goods and services are produced *without increases* of labor and capital. It is not about cutting costs. Productivity is "doing things right" and "doing the right things" to achieve maximum efficiency and value.

How can we increase productivity? Several measures can produce improvement. If you want to increase labor productivity, implement employee training and increase the use of technology so employees can produce more, or invest in equipment to help them work faster. In other words, productivity can be increased by improving all factors involved in production or improving only factors that affect productivity.

What are the key steps for increasing productivity? The first step is to establish areas that *need improvement by conducting* a productivity *study. The results* may reveal that you can increase sales by creating more innovative products and this may require redesigning products, jobs, or *processes*. The second important step is to measure and monitor progress *and* compare your results against industry benchmarks.

How can we improve labor productivity? Labor productivity is affected by various factors including (1) management competence; (2) knowledge, skills, and attitudes of workers; (3) levels of technology used; (4) effectiveness of processes and systems; and (5) investments in materials, processes, and equipment. Certain techniques can increase productivity, for example:

- Selling higher value goods
- Developing innovative new products
- Improving quality of existing products or services

- Reducing production costs by streamlining processes and producing larger volumes to achieve economies of scale
- Upgrading equipment
- Improving working environment
- Adopting new technologies to maximize capital

Why is productivity important? Economic growth cannot be sustained without *productivity* improvements. Europe's growth *depends* on the ability *of all industries* to maximize labor and capital use and implement effective work practices and innovation to achieve greater output.

9.1.2 Productivity terminology

In general, productivity signifies the measurement of how well a business entity uses its resources to produce outputs from inputs. Moving beyond this general notion, a glance at the productivity literature and its various applications quickly reveals no consensus as to the meaning of productivity and no universally accepted way to measure it. Attempts at productivity measurement have focused on individuals, firms, selected industrial sectors, and even entire economies. The intensity of debate about appropriate measurement methods appears to increase with the complexity of economic entity under analysis.

A number of productivity measures are used commonly. Choosing among them usually depends on the purpose of the productivity measurement and the availability of data. Productivity measures can broadly be placed into two categories. Single factor or partial productivity measures relate a particular measure of output to a single measure of input such as labor or capital. Total or multifactor productivity (MFPs) measures relate a particular measure of output to a group of inputs or inputs.

Productivity measures can also be distinguished by their reliance on a specific measure of gross output or on a value-added concept that attempts to capture the movements of outputs. Of the most frequently used MFP measures, capital–labor MFP relies on a value-added concept of output while capital labor–energy–materials (KLEMS) MFP is based on specific measures of gross output (OECD 2001). Based on OECD data, the five most widely used productivity concepts are:

Labor productivity based on gross output — This measurement traces the labor requirement per unit of output. It reflects change in the input coefficient of labor by industry and is useful for the analysis of specific industry labor requirements. Its main advantage as a productivity measure is its ease of measurement and readability, particularly because it requires only price indices on gross output. However, since labor productivity is a partial productivity measure, output typically reflects the joint influences of many factors.

Labor productivity based on value added — This technique is useful for analyzing micro–macro links such as an industry's contribution to economy-wide labor productivity and economic growth. Its main advantages are ease of measurement and readability, although the analysis does require price indices on intermediate inputs along with gross output data. In addition to its limitations as a partial productivity measure, value-added labor productivity presents several theoretical and practical drawbacks including the potential for double counting production of benefits and deflation (see next paragraph).

Capital–labor MFP based on value added — This productivity measure is also useful for the analysis of micro–macro links such as industry contributions to economy-wide MFP growth and living standards and to analyze structural changes. Its main advantage is the ease of aggregation across industries. The data for this measurement are also directly available from national databases. The main drawback is that this MFP is not a good measure of technology shifts at the industry or firm level.

Capital productivity based on value added — Changes in capital productivity denote the degree to which output growth can be achieved with lower costs in the form of foregone consumption. Its main advantage is its ease of readability but it suffers the same limitations as other partial productivity measurements.

KLEMS multifactor productivity — This measure is used to analyze technical industry-level and sector changes. It is the most appropriate tool to measure technical change by industry because it fully acknowledges the role of intermediate inputs in production.

See Table 9.1 for more information on productivity measures that are linked. For example, it is possible to identify various driving forces behind labor productivity growth, one of which is the rate of MFP change. This and other links between productivity measures can be established with the help of the economic theory of production.

9.1.3   Purposes of productivity measurement

A straight look at the productivity literature and its various applications reveals very quickly that productivity measurements do not fill a unique purpose and cannot be expressed as single measures. The objectives of productivity measurement include:

Technology—This is a very important measurement to determine productivity growth and is also known as tracing technical change. Technology has been described as "currently known ways of converting resources into outputs desired by the economy" (Griliches 1987) and is a vital factor in producing results or indicating advances in design and quality embodied in new products.

Efficiency—The quest for identifying efficiency changes is conceptually different from identifying technical change. Full efficiency in an

Table 9.1 Overview of Productivity Measures

	Input Measure			
	Single Factor Productivity Measures		Multifactor Productivity (MFP) Measures	
Output Measure	Labor	Capital	Capital and Labor	Capital, Labor, and Intermediate Inputs (Energy, Materials, and Services)
Gross output	Labor productivity based on gross output	Capital productivity (based on gross output)	Capital and labor MFP based on gross output	KLEMS multifactor productivity
Value added	Labor productivity based on value added	Capital productivity (based on value added)	Capital and labor MFP based on value added	–

engineering sense means that a production process achieves the maximum output that is physically possible with current technology and a fixed number of inputs (Diewert and Lawrence 1999). When productivity measurement meets industry level, efficiency gains can be accomplished by improved efficiency in an individual operation or by a shift of production toward more efficient outputs.

Real cost savings—Even if it is possible to improve and enhance technology, real cost savings remains a challenge for all industries and difficult to achieve at an operations level.

Benchmarking production processes—In business economics, comparisons of productivity measures for specific production processes can help identify inefficiencies.

Living standards—Measurement of productivity is an important element for assessing standards of living. A simple example is per capita income (probably the most common measure of living standards). Income per person in any economy varies directly with one measure of labor productivity: value added per hour worked.

9.1.4 Space optimization

The value of space is an important resource that is sometimes overlooked by manufacturing operations. Some companies that own the spaces they occupy do not view the spaces as representing significant value. A result is that the operations expand into available space without planning for optimal use. Potentially valuable floor space is used for equipment grave-yards or storage of outdated products. Unused buildings are allowed to deteriorate and lose value although they remain taxable and continue to accrue insurance costs.

Several options can be explored in an effort to improve returns from unoptimized space: (1) optimize existing floor layouts, (2) lease excess space to outside interests, (3) reduce leased space requirements, (4) avoid new construction to accommodate expansion; and (5) demolish deteriorating structures to reduce tax and insurance expenses.

Other possible savings opportunities to consider are reducing operating costs for the space (lighting, heating, ventilation, air conditioning, other utilities) and decreasing transportation costs, for example by consolidating operations to eliminate traveling between two sites.

9.1.5 Productivity enhancement: Increasing work per person per hour

A number of techniques that can help enhance productivity are described below.

9.1.5.1 Machine setup

The preparation of a machine to perform a specific task or operation is called setup time. It is possible to divide setup into internal and external tasks. Internal setup requires an interruption in production. For that reason it is critical to minimize the time required for internal setup. External setup time is used to prepare machines and tooling for the next operation without stopping production. As a result, it is not as essential to minimize the time required.

Changing from internal to external setup procedures usually cuts required time by 30 to 50%. The first step is to identify internal and external setup tasks. Informal discussions with workers and observation of the process often suffice. The second step is to convert as many internal tasks as possible into external tasks. The third segment involves the streamlining of all steps to reduce the required times. Operators must be involved because they participate in the process at their work locations.

9.1.5.2 Bottleneck mitigation

A bottleneck in manufacturing is a process or procedure that restricts the entire operation from running faster. It is usually identified by idle workers or machines. Production time is lost while they wait for the bottleneck to dissipate. Identifying and alleviating bottlenecks in a manufacturing process leads to increases in throughput (pieces produced per person per hour) of a process or even an entire plant. In most cases, increased throughput will increase revenue; however, it could also reduce production lead time.

To identify a bottleneck, the first step is interviewing relevant workers and supervisors. However, two cautions are noted. First, many companies do not consider the effect of net production rate including scrap or waste on a bottleneck. They may not define a process or procedure as a bottleneck if a defective product can be re-used as a raw material as is possible in the paper, glass, plastics, and metals industries. Second, it is important to realize that a perceived (or theoretical) bottleneck may not, in fact, be the real problem. This is why discussions about bottlenecks should involve machine operators.

Alleviating bottlenecks is often accomplished by improved equipment or tools. It can also be accomplished by the purchase of manufactured parts, by parameter controls (temperature and/or humidity), or better quality raw materials. Increasing production may involve additional costs for more or better raw materials, labor, and energy. On a per-piece basis, however, these quantities should remain constant or even decrease in some cases. Industries that have high internal rejection rates may not require more raw materials if their defect rates are reduced. If alleviating a bottleneck also reduces re-work, more labor may not be required.

9.1.5.3 Defect reduction

Defect reduction should be a focal point when an industrial enterprise considers potential productivity improvements. Time spent producing defective parts is costly. Labor, energy, and raw materials are wasted. Increased costs are incurred for inspection, inventory operations, material handling, and clean-up.

One shortcoming of many statistic-based quality control systems (inspections are performed at the end of the production line) is that they are intended only to separate defective products from good ones. While this effectively reduces the number of defective products that are passed along to customers, the system does almost nothing to reduce the number of defective items produced. The time between defect detection and feedback to the process line is often hours or days—too long to be helpful to the business.

Often, the most difficult task in identifying and estimating the potential benefits of an assessment recommendation is obtaining reliable information. Fortunately, the nature, frequency, and causes of product defects are normally well known by quality control personnel and machine operators. The major tasks involved in a defect reduction recommendation are (1) identifying the most costly defects; (2) determining why they occur; and (3) recommending actions to reduce or eliminate the defects.

After the causes of the most costly defects are identified, installing new sensors may be a solution or at least part of a solution. Sensors can detect defects and, more importantly, provide valuable feedback to machine operators or directly to the process equipment. Defects should be detected as soon as possible after they occur to avoid adding value-added processes to a product that is already defective. In fact, sometimes a defect can be detected before it occurs. If the causes are known (e.g., temperature, humidity, pressure, movement, positioning, tool defects), sensors can be installed to detect when those conditions occur, adjust the appropriate parameters or stop the process before defects are produced.

Immediate feedback can also be used to prevent serial defects (batches of defective products) by shutting down the production line until the cause is identified and corrected or by indicating to the operators that parameter adjustments (such as changing temperature or pressure) are needed. In cases where serial defects are common, shutting down the process or machine is recommended unless re-start requirements are extensive. When defects occur as abnormalities or isolated incidents, it is usually best to automatically reject the defective part and scrap, recycle, or repair it rather than shut down the line.

9.1.5.4   Labor optimization

Labor cost saving opportunities arise when worker time or skills are underutilized. Many factories maintain operations that have remained unchanged for decades. Workers perform better when they use automated equipment for inspecting and packaging, for example. Some workers must move materials or products over excessive distances, wasting time and money. Ineffective production creates a disinterested, discouraged, and inefficient workforce. Workers perform less than ideally when their efforts exert little or no effect on the result.

Conditions that increase labor cost without enhancing productivity are unclear definitions of defects, duplicative inspection procedures, and equipment that requires constant adjustments. However, workers must be trained adequately to operate new technologies. The development of standard operating procedures will allow manufacturers to demystify production processes and train or cross-train workers to perform needed operations efficiently.

9.1.6   Productivity enhancement: Decreasing per-piece cost

9.1.6.1   Scheduling

One type of scheduling recommendation involves adequate planning when certain types of changeovers are involved, for example, in painting or printing operations. Better sequencing of runs from light to dark color can eliminate cleaning operations during color changes. Paint guns must be cleaned regularly in painting operations and cylinder blankets and spray jets must be cleaned in printing houses. Since lighter colors are affected by darker ones to a large degree, the cleaning operations must be performed thoroughly and complete or partial elimination of cleaning tasks would lead to significant savings.

Another example of a beneficial schedule change on a production is the elimination of waiting for batches of products. If waiting is eliminated during all sequential operations for one part or product, precious time and money resources are saved. In some cases, simple reorganization of work schedules can lead to improvements.

Scheduling and Just-in-Time (JIT) strategies are closely related. The purpose of scheduling of all production runs is to produce parts as economically as possible, maintain somewhat constant workloads, and deliver parts to their destinations quickly. Effective scheduling of production runs to maintain constant throughput is a complex matter that requires more than a one-day assessment. Evaluating a production system and developing recommendations are tasks for an experienced assessor. Certain manufacturing processes require long production times (e.g., when a product undergoes annealing, melting, or needs specific exposure times) during which a product is not worked on actively. Times required for such inactive processes must be included in scheduling modifications.

9.1.6.2 Purchasing

Improvements in raw material and component purchasing can lead to significantly reduced material costs, in-house material handling, and inventory carrying costs. During a plant assessment, raw material and purchased component inventories should be surveyed and close attention paid to in-house labor efforts associated with handling materials. If existing delivery schedules require significant handling of raw materials and provision of adequate floor space, it may be profitable to switch to a JIT delivery schedule.

9.1.6.3 Burden analysis

Indirect expenses of a manufacturing organization are known as burden and overhead. A wide variety of items collectively constitute burden or overhead:

- Salaries and wages of foremen, inspectors, clerical employees, crane operators, etc.
- Purchasing, receiving, and shipping
- Utilities (heat, light, water, etc.) and maintenance
- Taxes, insurance, rentals, and depreciation
- Administration (sales, technical, management) and marketing

The selling price of a manufactured item consists of:

- Direct costs of labor and materials
- Factory overhead
- General administrative and marketing costs
- Profit

In cost accounting, overhead includes factory, general administrative, and marketing costs (and not direct labor, materials, and profit). Some overhead expenses vary with production rates and some do not. This is why it is difficult to assign a truly accurate value of burden to each item produced.

9.1.6.4   *Inventory*

Industrial facilities normally maintain four types of inventories: (1) raw materials, (2) purchased components, (3) work in progress (WIP), and (4) finished goods. Each type of inventory involves carrying costs for: (1) investments (time value of money tied up in inventoried goods), (2) damage or spoilage, (3) product obsolescence, (4) storage space, and (5) in-house handling and transportation.

Although the magnitudes of such costs may vary among manufacturers, it is easy to recognize that all such costs will increase over time. In fact, if inventory retention periods are long enough, the carrying costs can exceed the profits gained from the eventual sale of the products and a net loss will result. For this reason, it is desirable to minimize inventories of all types.

9.1.6.5   *Floor layout*

Modifications of a floor layout may save time, space, inventory or labor by eliminating or reducing wasted motion or non-value added activities. A common practice by plant management over the years has been to expand into all available space, even constructing or leasing new buildings rather than maximizing existing floor layouts. Most manufacturers did not design the layouts you see today, but rather they "grew" into them. Typically, manufacturers do not assign any value to wasted space, particularly if it does not represent a direct cost.

Floor layouts vary based on number of products made and quantities produced. Low quantity producers, or job shops, produce small quantities of specialized products using general purpose equipment and skilled workforces. One type of small quantity layout is process oriented. The facility is separated into departments and equipment is consolidated by type, for example, lathes in one area and paint booths in another. The product is routed through departments based on the order of the processes required. This layout allows for flexibility but presents the disadvantage of requiring much moving and handling. If the product varieties are few and the demand is predictable, the product can be produced in batches, characterized by longer production runs that are repeated frequently. Equipment is somewhat specialized, and production is intended to replenish inventory since equipment changeover is time consuming.

High quantity production is generally set up in a flow line layout. Product moves through the factory, usually on conveyors, and the workstations and workers are located next to the line. Many factories include a mixture of these types of layouts.

9.2   Productivity growth and energy consumption in Romania

Productivity growth depends on real-life needs and economic development of every country or state. Energy consumption is proportional to industrialization and economic development. Productivity growth and energy consumption are closely related in industry and in domestic life. This section presents data from Romania along with information about gross domestic product (GDP) variations in recent years. Also, data from other European countries will be presented as a base for comparison.

Romanian economic growth is among the fastest in the European Union (EU; World Bank 2006). It is the 11th largest economy in the EU based on total nominal GDP and the 8th largest based on purchasing power parity. Romania is one of the fastest growing markets in recent history with consistent annual gross national production (GNP) growth rates above 8% (+36.18% for 2007) since 2000. Romania is the 7th largest member state of the EU and its most important trading partner. Its capital, Bucharest, is one of the largest financial centers in the region, with a metropolitan area of more than 2.6 million people.

Romania has experienced growth in foreign investment with a cumulative foreign direct investment (FDI) totaling more than $100 billion since 1989 (*Economist* 2008).

Some economic predictions indicated that Romanian GDP would double from 2006 to 2011 (Diaconu et al. 2009). Preliminary estimates for 2009 showed real GDP growth of −4%, while the forecasts for 2009 and 2010 project an average of 6 to 6.5% per year (European Commission 2006).

These predictions of GDP growth were affected by economic problems in recent years. Before the recession in 2009, some economic predictions indicated that Romanian GDP would double from 2006 to 2011. One scholar even suggested that Romania would overtake Italy in GDP per capita by 2020. Despite a growth rate of 7.1% in 2008, the Romanian economy was heavily affected by the global financial downturn and contracted by some −7.2% in 2009. The European Council and independent

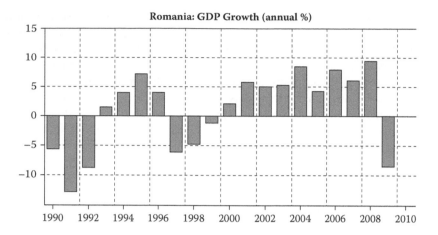

Figure 9.1 Romanian GDP growth from 1990 to 2010. (*Source:* World Bank. www. worldbank.org.ro)

analysts predicted that growth would resume in 2010. Future prospects are tied to the country's increasingly important integration with the EU member states. The country is expected to join the Eurozone in 2014.

The Romanian economy has undergone a transformation from agricultural to industrial as a result of rapid urbanization (Figure 9.1). The country's energy consumption has risen rapidly as a result of social and economic development. In recent years, the average increase in domestic electricity consumption has been 3.5% per year, and it is projected to maintain a similar rate in the coming years despite the economic crisis starting in 2009 mainly because of domestic consumer needs.

This study intends to be an overview of energy policies in Romania since 2006, with a particular accent on electricity demand in conjunction with economic growth.

9.2.1   Economic background

A government commitment to sound economic policies since 2001 placed the Romanian economy in a good position to embark on a sustained path of faster economic growth (Robu et al. 2007). Between 2000 and 2008, real GNP grew by 450% on a cumulative basis and 21.6% on average, making the Romanian economy one of the fastest growing in the world.

Romania's GNP has grown at an average annual rate of 16.36% since 2000 although the pattern has been uneven. As shown in Table 9.2, the economy rebounded with a maximum of 36.18% GNP growth in 2007. Romania was one of the world's fastest-growing countries after 2002, a

Table 9.2 Romania's Gross National Production (GNP)
from 2000 through 2010

Year	GNP (Billions of Dollars)	GNP Growth (%)
2000	37	3.93
2001	40.1	8.38
2002	45.8	14.21
2003	56.9	24.24
2004	73.1	28.47
2005	98.6	34.88
2006	121.9	23.63
2007	166	36.18
2008	200	20.48
2009	161.1	−19.45%
2010	161.7	0.37%

year in which GNP increased 14.21% (Robu et al. 2007) and reached U.S. $45.8 billion.

After the historical levels recorded in 2002, GNP rose every year, far beyond the targets and forecasts. GNP grew from U.S. $37 billion in 2000 to U.S. $200 billion. In 2007, Romania reached the highest economic growth among the members of United Nations Organisation for Economic Cooperation and Development (OECD) and the second highest in the EU (World Bank 2006). Romania's GNP from 2009 through 2015 is expected to increase by U.S. $56 billion per year.

Exports in 2008 amounted to U.S.$52.2 billion, breaking all records in the history of the country (U.S. Central Intelligence Agency 2009) and placing Romania at the 56th position in world exports. Economic growth in recent years has been associated with the privatization of public enterprises and productivity growth.

9.2.2 Productivity growth by GDP and energy consumption

The EU Sustainable Development Strategy (EU SDS) sets out the objectives of promoting sustainable consumption and production patterns and positively impacting productivity. Addressing social and economic development within the carrying capacities of ecosystems and decoupling economic growth from environmental degradation are essential requirements for sustainable development. The headline indicator is productivity with established targets in electricity and energy consumption among other indicators like waste generation (European Union n.d.). Figure 9.2

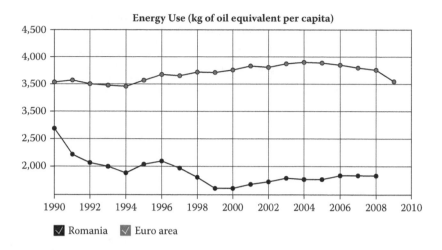

Figure 9.2 Energy use in Romania and EU.

compares energy use in Romania and the EU. European statistical (Eurostat) data are presented in Tables 9.3 through 9.5.

Table 9.3 depicts fuel-based energy supplied to final consumers for all uses (industry, transport, households, services, agriculture, etc.). Final energy consumption in industry covers use in all industrial sectors except for energy production. The fuel quantities transformed in the electrical power stations of automobile producers and the quantities of coke transformed into blast furnace gas are not part of the overall industrial consumption and belong to the transformation sector. Final energy consumption in transport covers consumption by rail, highway, air, and inland navigation transportation. Final energy consumption in households covers quantities consumed by private households, commercial enterprises, public administration, services, agriculture, and fisheries. Table 9.4 shows energy losses from fuel-based operations.

Based on Table 9.5, resource productivity equals GDP divided by domestic material consumption (DMC). DMC measures the total materials directly used by an economy. It is defined as the annual quantity of raw materials extracted from the domestic territory plus all physical imports minus all physical exports. It is important to note that the *consumption* term in the DMC context denotes apparent (not final) levels. DMC does not include upstream flows related to imports and exports of raw materials or products originating outside the focal economy. When examining resource productivity trends over time in a single geographic region, the GDP should be expressed as Euros in chain-linked volumes to reference year 2000 at 2000 exchange rates.

Table 9.3 Industry Energy Consumption, 2000 to 2009[a]

Country	Year									
	2000	2001	2002	2003	2004	2005	2006	2007	2008	2009
Belgium	14059	14090	12712	12925	12559	11711	12405	12198	11928	9614
Bulgaria	3653	3651	3491	3817	3774	3799	3889	3930	3485	2430
Czech Republic	10119	9752	9582	9630	10019	9682	9718	9451	9007	8116
Denmark	2932	3024	2845	2859	2896	2863	2907	2821	2692	2329
Germany	57553	56540	56554	63408	62570	62488	62185	61747	60601	51794
Estonia	571	631	575	673	687	719	692	771	755	541
Ireland	2497	2469	2415	2403	2526	2631	2836	2590	2520	2161
Greece	4447	4504	4442	4327	4068	4158	4231	4601	4209	3462
Spain	25372	27130	27454	29325	30117	30977	25714	27823	26423	23790
France	37171	39694	38655	39275	37839	35799	36006	35211	36018	28993
Italy	39737	38768	38707	40744	39919	39624	38362	37690	36577	29546
Cyprus	441	426	423	442	446	319	279	283	306	260
Latvia	576	613	622	626	667	699	741	723	679	652
Lithuania	780	772	863	909	937	995	1055	1064	956	821
Luxembourg	715	715	706	730	826	749	815	775	753	617
Hungary	3513	3618	3699	3591	3347	3374	3385	3346	3342	2672
Malta	43	42	44	48	47	42	46	46	48	75
Netherlands	14829	14662	14606	14739	14998	15506	13441	13034	12659	12854
Austria	7246	7408	7409	7798	8121	8819	8763	8973	9014	8263
Poland	18984	17503	16785	17374	17983	16593	17021	17829	16368	14730
Portugal	6293	6171	6269	5846	5870	5868	5831	5899	5572	5175
Romania	9110	9634	10319	10130	9983	9942	9579	9139	8798	6411
Slovenia	1423	1335	1262	1498	1549	1643	1698	1604	1483	1230
Slovakia	4532	4491	4686	4860	4598	4696	4791	4607	4535	4052
Finland	11803	11474	12220	12699	13059	12157	13270	12957	12318	10120
Sweden	14264	13201	13196	12827	12942	12628	12661	12795	12288	11155
United Kingdom	36665	36294	34724	35354	34181	34325	33724	32830	32537	27594
Norway	7011	6554	6152	6609	6940	6821	6648	6604	6723	5635
Switzerland	3874	4043	3867	3944	4016	4060	4130	4070	4099	3811
Croatia	1396	1450	1401	1451	1571	1581	1641	1662	1698	1427
Turkey	20985	16707	19904	22482	22817	22718	25374	25943	20744	20406

[a] Expressed as 1000 tonnes of oil equivalent (toe).

Table 9.3 reveals decreasing of industrial energy consumptions of a few countries including Romania, starting 2009 when the economical crisis started in Europe. Table 9.4 shows slight decreases of residential energy consumption in a few countries including Romania, also starting in 2009. Romanian energy consumption started to increase in 2001 due to increased economic development.

Table 9.4 Residential Energy Consumption, 2000 to 2009[a]

Country	Year									
	2000	2001	2002	2003	2004	2005	2006	2007	2008	2009
Belgium	9474	9874	9332	9866	10017	9920	8914	8107	8778	8300
Bulgaria	2155	2003	2164	2284	2117	2117	2167	2068	2117	2116
Czech Republic	6023	6562	6029	6313	6235	6216	6479	6043	5990	5984
Denmark	4133	4404	4302	4400	4400	4447	4452	4494	4478	4456
Germany	65188	69671	67137	63676	63246	63656	64805	61383	68160	65786
Estonia	929	949	918	926	923	890	882	963	953	966
Ireland	2505	2640	2627	2733	2836	2908	3068	2900	3157	3070
Greece	4486	4702	4898	5488	5399	5497	5490	5377	5212	4848
Spain	11882	12479	12808	13784	14591	15061	15806	15872	15801	14887
France	45251	42590	40642	42219	44215	43395	42764	39763	42574	44616
Italy	28133	29421	28317	29897	31002	31232	29355	27196	27273	28677
Cyprus	174	172	191	207	199	314	287	298	294	311
Latvia	1327	1442	1431	1499	1473	1504	1481	1458	1452	1517
Lithuania	1343	1373	1378	1385	1379	1388	1434	1356	1381	1379
Luxembourg	475	511	494	516	551	544	538	490	527	567
Hungary	5587	6002	6009	6596	6089	6458	6208	5554	5570	5520
Malta	76	74	80	88	88	76	81	81	80	68
Netherlands	10299	10658	10262	10522	10479	10143	10062	9300	9862	10190
Austria	6131	6390	6269	6348	6214	6690	6351	5997	6136	6161
Poland	17191	18795	17766	17752	17816	18343	19251	18394	18618	18738
Portugal	2804	2859	2987	3115	3217	3224	3219	3226	3121	3204
Romania	8408	7278	7217	7819	7965	7990	7854	7518	8070	8015
Slovenia	1125	1119	1162	1249	1239	1186	1158	1048	1115	1092
Slovakia	2586	3082	2998	2840	2666	2540	2310	2081	2131	2147
Finland	4539	4913	5021	5108	4981	5015	5127	5148	5036	5367
Sweden	7294	7506	7331	7378	7144	7302	7002	6730	6637	6949
United Kingdom	43033	44276	43230	43859	44753	44151	43015	41502	42497	40275
Norway	3825	3984	3978	3812	3740	3874	3814	3839	3836	3992
Switzerland	5603	5853	5717	6034	6067	6227	6060	5592	5998	5929
Croatia	1665	1666	1730	1873	1887	1927	1859	1722	1786	1809
Turkey	17595	16304	16655	17501	18074	19309	19894	20729	22608	20529

[a] Expressed as 1000 tonnes of oil equivalent (toe).

9.3 Productivity growth via energy savings in industrial heating sector

Energy efficiency is generally the largest, least expensive, most quickly deployable, least visible, misunderstood, and neglected way to enhance energy services. The largest energy user in most countries is industry; about half of all industrial energy use powers specific processes in

Table 9.5 Resource Productivity (GDP/DMC) for 2000 through 2007[a]

Country	Year							
	2007	2006	2005	2004	2003	2002	2001	2000
Belgium	1.47	1.43	1.43	1.43	1.41	1.36	1.29	1.32
Bulgaria	0.14	0.13	0.14	0.13	0.13	0.13	0.13	0.13
Czech Republic	0.42	0.4	0.39	0.36	0.37	0.37	0.34	0.33
Denmark	1.24	1.2	1.22	1.31	1.36	1.38	1.32	1.28
Germany	1.71	1.65	1.64	1.58	1.58	1.55	1.52	1.41
Estonia	0.27	0.31	0.31	0.28	0.25	0.32	0.34	0.32
Ireland	0.66	0.66	0.68	0.68	0.67	0.67	0.63	0.63
Greece	0.98	0.94	0.9	0.87	0.83	0.86	0.86	0.88
Spain	0.9	0.85	0.87	0.86	0.85	0.87	0.93	0.93
France	1.8	1.83	1.83	1.74	1.87	1.74	1.73	1.64
Italy	1.6	1.52	1.49	1.52	1.62	1.46	1.36	1.25
Cyprus	0.64	0.66	0.62	0.61	0.67	0.61	0.66	0.66
Latvia	0.31	0.31	0.29	0.29	0.29	0.27	0.27	0.24
Lithuania	0.43	0.46	0.43	0.42	0.38	0.47	0.53	0.44
Luxembourg	4.32	3.04	3.33	3.33	3.01	2.95	3.01	2.78
Hungary	0.6	0.47	0.38	0.42	0.46	0.45	0.43	0.45
Malta	2.14	2.18	2.42	2.32	2.81	3.04	3.26	3
Netherlands	2.6	2.59	2.45	2.42	2.44	2.35	2.14	2.16
Austria	1.4	1.32	1.3	1.3	1.37	1.38	1.44	1.41
Poland	0.38	0.4	0.38	0.37	0.38	0.38	0.35	0.32
Portugal	0.62	0.62	0.7	0.7	0.75	0.68	0.64	0.66
Romania	0.14	0.16	0.16	0.16	0.16	0.17	0.15	0.18
Slovenia	0.46	0.48	0.53	0.49	0.5	0.51	0.51	0.48
Slovakia	0.49	0.44	0.39	0.4	0.43	0.4	0.39	0.4
Finland	0.79	0.78	0.8	0.79	0.76	0.78	0.76	0.76
Sweden	1.79	1.95	1.69	1.85	1.83	1.81	1.82	1.71
United Kingdom	2.54	2.48	2.41	2.28	2.3	2.24	2.13	2.11
Norway	1.23	1.26	1.22	1.16	1.2	1.22	1.18	1.13
Switzerland	3.36	3.18	3.14	3.17	3.24	3.15	3.09	3.05

[a] Expressed as euros per kilogram at 2000 exchange rates.

energy-intensive industries like heating (Minea 2008). Furnaces are used in a wide variety of applications including power plants, nuclear reactors, refrigeration and heating systems, automotive production, heat recovery systems, chemical processing, and food industries (Minea 2008, 2010a and b).

One major consumer of industrial energy is heating equipment such as high temperature heat treatment and forging furnaces used in industrial processes. Furnaces are used for industrial purposes world-wide and represent an annual energy consumption exceeding a billion joules. Their domain of use ranges from food and chemical production to metallurgical industries. Furnaces were used in the earliest civilizations. Because of their wide variety of applications, they are still key devices in modern industries. Electric and combustion-heated furnaces are the most widely used and serve as the backbones of most infrastructures.

The performance improvement of heating equipment must be corre-lated with energy consumption and consumption is mainly reflected by production price. The need to reduce energy consumption and environ-mental footprints places high demand on redesigning the inner spaces of heating devices. Modifications of chamber geometry can potentially lead to energy savings and even reduce furnace production costs. Studies and industrial acceptance already note such potential; however, the studies were largely based on smaller scale furnaces (Minea 2010). Industry faces a pressing need to expand research to study large industrial scale heating processes and deliver novel designs that will significantly reduce energy consumption and environmental footprints.

9.3.1   Energy savings and performance improvements: Fuel-based systems

Figure 9.3 shows a typical fuel-based process heating system and poten-tial for improving system performance and efficiency. Most of the

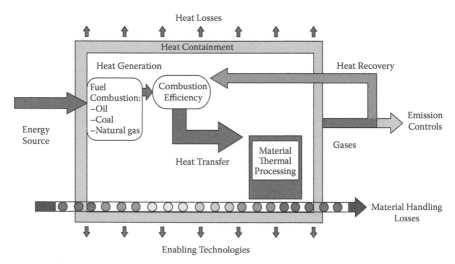

Figure 9.3 Fuel-based process heating system and opportunities for improve-ment. (From Minea, A.A. 2013. In *Advances in Industrial Heat Transfer,* Minea, A.A. (Ed.), Boca Raton, FL: Taylor & Francis. With permission.)

opportunities are interdependent, for example, the dependence of heat recovery on heat generation. Transferring heat from exhaust gases to incoming combustion air reduces system energy losses and also promotes more efficient fuel combustion, thereby delivering more thermal energy.

Fuel-based process heating equipment is used by industry to heat materials under controlled conditions. Recognizing opportunities and implementing improvements are the most cost-effective processes when they combine a systems approach with awareness of efficiency and performance improvement opportunities that are common to systems utilizing similar operations and equipment.

It is important to recognize that a particular type of process heating equipment can serve different applications and that a particular application can be served by a variety of equipment types. For example, the same type of direct-fired batch furnace can be used to cure coatings on metal parts at a foundry and heat-treat glass products at a glassware factory. Similarly, coatings can be cured in batch-type or continuous furnaces. Many performance improvement opportunities are applicable to a wide range of process heating systems, applications, and equipment. This section provides an overview of basic characteristics to identify common components and classify systems. Equipment characteristics determine the most likely opportunities for application of performance and efficiency improvements.

Heating methods — In principle, we can distinguish direct and indirect heating methods. Systems using direct heating methods expose the material to be treated directly to a heat source or combustion product. Indirect heating methods separate the heat source from the load and may use air, gases, or fluids as media to transfer heat from the heating element to the load. Convection furnaces utilize indirect heating.

Heating element — Among the many types of basic heating elements that can be used in process heating systems are burners, radiant burner tubes, heating panels, and bands.

9.3.1.1 Efficiency enhancement opportunities
The remainder of this section gives an overview of the most common performance improvement opportunities for fuel-based process heating systems (Minea 2013). The performance and efficiency of a process heating system can be described with an energy loss diagram (Figure 9.4). The main goals of performance optimization are reduction of energy losses and increase of energy transferred to loads. It is therefore important to know which aspects of a heating process deliver the greatest impact. Some of the principles discussed here also apply to electric- or steam-based process heating systems (Minea 2013).

Performance and efficiency improvement opportunities can be grouped into four categories as shown in Figure 9.3: (1) heat generation,

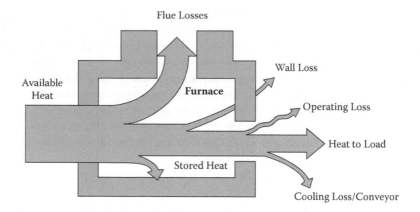

Figure 9.4 Energy loss in fuel-based process heating system. (From Minea, A.A. 2013. In *Advances in Industrial Heat Transfer*, Minea, A.A. (Ed.), Boca Raton, FL: Taylor & Francis. With permission.)

in particular, the equipment and the fuels used to heat a product; (2) heat containment involving methods and materials that can reduce energy loss to the surroundings; (3) heat transfer intended to increase heat transferred to loads, reduce energy consumption, increase productivity, and/or improve quality; (4) enabling technologies that reduce energy losses by improving material handling practices, effectively sequencing and scheduling heating tasks, seeking more efficient process control, and improving the performances of auxiliary systems.

Figure 9.4 shows several key areas where the performance and efficiency of a system can be improved. It is important to note that many opportunities affect multiple areas. Transferring heat from exhaust gases to incoming combustion air or cold process fluid reduces energy losses from a system and allows delivery of more thermal energy to heated material, thus improving fuel efficiency.

9.3.2   Energy savings and performance improvements: Electric-based systems

Electric-based process heating technologies in manufacturing operations use electricity to make or transform a product through heat-related processes. These systems perform operations such as heating, drying, curing, melting, and forming (Minea 2013).

Electric-based process heating systems are controllable, clean, and efficient. In some cases, these technologies are chosen for unique technical capabilities; in other cases, the lower cost of electricity in relation to the cost of natural gas or other fuel is the deciding factor. Certain operations cannot be performed economically without electric-based systems.

For some industrial applications, electric-based technologies are used throughout operations. Alternatively, electric-based systems are used only in certain niche applications.

9.3.2.1 Types of electric-based systems

Electric-based process heating systems use electric currents or electromagnetic fields to heat materials. Direct heating methods generate heat within a workpiece by (1) passing an electrical current through the material, (2) an electrical (eddy) current into the material, or (3) exciting atoms and/or molecules in the material with electromagnetic radiation (e.g., microwaves). Indirect heating utilizes one of the three methods to heat an element or susceptor that transfers the heat to the workpiece via conduction, convection, radiation, or a combination. The remainder of this section covers these process heating electrotechnologies:

- Arc furnaces
- Infrared electric processing
- Electron beam processing
- Induction heating and melting
- Laser heating
- Microwave processing
- Arc and non-transferred arc plasma processing
- Radiofrequency processing
- Direct and indirect resistance heating and melting
- Ultraviolet curing

In industrial heating, resistance is the most widely used technique due to its clear advantages as was stated earlier. Improving the efficiency of electric based-systems can lead to productivity enhancement. A few considerations about resistance heating will be explained.

9.3.2.2 Improving efficiency of existing resistance heating systems

A few well-known methods can improve the efficiency of resistance heating, for example:

- Better process control systems, including those with feedback loops, use less energy per product produced. Good control systems allow precise application of heat at the proper temperature for the correct time.
- Clean resistive heating elements can improve heat transfer and process efficiency.
- For insulated systems, improvements in heat containment can reduce losses of energy to the surroundings.
- Match heating elements closely to the geometry of parts to be heated can save energy.

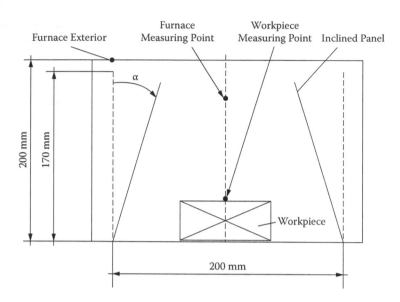

Figure 9.5 A method to save energy in electrical furnaces: Experimental setup.

New methods intended to save energy can be more challenging. One new method is redesign of heat chamber geometry to enhance heat transfer inside the chamber. Besides improving furnace performance, heat transfer enhancement allows exterior furnace size to be increased considerably.

Simplified models based on simulation, empirical, and experimental methods can estimate heat transfer enhancement in industrial processes (Minea 2010c, 2009a and b). Some investigations were done to examine radiation enhancement when panels were introduced inside a furnace (Figure 9.5). The intent was to explore the suitabilities of different radiation panels for semi-industrial heating of commercial parts. A positive result would produce immense benefits from environmental and economical views. The work was carried out with the following objectives:

- Evaluation of various widths of radiant panels
- Evaluation of various radiation surfaces
- Comparison of energy consumption levels
- Assessment of heating behavior based on panel position, width, and surface variations

Experimental results suggested that the optimal panel inclination was about 15 degrees. Also, a 22.5% energy saving (compared to a control experiment with no panels) was noted.

Implementing these measures can lead to furnace energy savings in one of two ways. Maximizing the rate of heat transfer can reduce required heating periods, thus decreasing energy consumption and lowering running costs. Alternatively, a more modest heating system at lower capital cost could be just as effective.

However, it must be noted that increasing the spacing of radiant panels may require a larger furnace chamber or treatment of smaller loads. These limitations should be assessed when proposed designs are considered. The velocity correction factor also requires improvement to meet a wider range of furnace conditions.

References

Diaconu, O., Oprescu, G.. and Pittman, R. 2009. Electricity reform in Romania, *Utilities Policy,* 17, 114–124.

Diewert, E.W. and Lawrence, D. 1999. Measuring New Zealand's Productivity. Treasury Working Paper 99/5. http://www.treasury.govt.nz/workingpapers/99-5.htm

Economist. 2008. Romania outlook, 2008–2009. www.economist.com

European Commission. 2006. Monitoring Report on the State of Preparedness for EU Membership of Bulgaria and Romania. Brussels.

European Union. n.d. http://epp.eurostat.ec.europa.eu

Griliches, Z. 1987. Productivity: Measurement Problems. In J. Eatwell, M. Milgate and P. Newman (Eds.), *The New Palgrave: A Dictionary of Economics.* New York: Stockson Press.

Minea, A.A. 2013. Heat transfer enhancement in process heating. In Minea, A.A., Ed., *Advances in Industrial Heat Transfer.* Boca Raton, FL: Taylor & Francis.

Minea, A.A. 2010a. A Mach number simulation study on a regular furnace. *Metalurgia International,* 15, 10–15.

Minea, A.A. 2010b. An Experimental method to decrease heating time in a commercial furnace. *Experimental Heat Transfer,* 23, 175–184.

Minea, A.A. 2010c. Simulation of heat transfer processes in an unconventional furnace. *Journal of Engineering Thermophysics,* 19, 31–38.

Minea, A.A. 2009a. CFD study on heat transfer in a muffle furnace. *International Review of Mechanical Engineering,* 3, 319–325.

Minea, A.A. 2009b. Experimental technique for saving energy in oval furnaces. *Environmental Engineering and Management Journal,* 8, 463–468.

Minea, A.A. 2008. Experimental technique for increasing heating rate in oval furnaces. *Metalurgia International,* 13, 31–35.

OECD. 2001. *Productivity Manual: A Guide to the Measurement of Industry-Level and Aggregate Productivity.* Geneva: United Nations Organisation for Economic Cooperation and Development. oecd.org/subject/growth/prod-manual.pdf

Robu, V., Serban, E.C., and Badileanu, M. 2007. The analysis of the reaction of the Romanian companies supplying electrical energy to the modification of the geopolitical context and of the internal legislative frame. *Theoretical and Applied Economics,* 1, 3–12.

U.S. Central Intelligence Agency. 2009. *World Factbook: Romania.* https://www.cia.gov/library/publications/the-world-factbook/geos/ro.html#Econ

World Bank. 2006. *Romania: Country Economic Memorandum. Promoting Sustained Growth and Convergence with the European Union*, Vol. II. Washington: World Bank Group. www.worldbank.org.ro

chapter ten

Productivity in healthcare organizations

Priyadarshini Pennathur, Arunkumar Pennathur,
Paulina Cano, Natalia Espino, and Jacqueline Loweree

Contents

10.1   Introduction

The healthcare industry has recently been in the spotlight in the United States. Managing resources and enhancing the productivity, efficiency, and effectiveness of healthcare work continues to pose significant challenges for this multibillion-dollar industry. With the increased emphasis on the design and use of healthcare information technology and electronic health records, demands on the skills and resourcefulness of healthcare workers to improve healthcare delivery have multiplied. Workers must continue to provide excellent care while retooling to use information technology, find ways to reduce healthcare delivery costs, and be productive and efficient in handling increasing numbers of patients.

Does the quantity of patients compromise quality of care? Does health information technology increase productivity or create an additional administrative burden on providers and actually lower their productivity? Do healthcare performance measurement and evaluation systems reward quality care or focus only on the number of patients seen? This book chapter explores such questions and discusses the major factors influencing productivity in healthcare organizations.

The chapter is organized into sections. Section 10.2 is a brief review of macrolevel international healthcare productivity and efficiency data. Section 10.3 covers the definitions of productivity from a traditional industrial perspective and highlights important issues in healthcare productivity assessment. Section 10.4 discusses the factors affecting productivity in healthcare. Metrics and tools used to measure productivity are also highlighted.

10.2   International healthcare productivity data

A country's healthcare system and levels of productivity can be described in terms of inputs and outputs. The inputs of a healthcare system are capital investments, government spending, and other elements. Outputs are patient satisfaction, efficiency, and productivity.

Data from the United Nations Organisation for Economic Cooperation and Development (OECD 2011) demonstrate how the United States ranks low in outputs for most macroindices. Ironically, U.S. input investments are among the highest. For example, according to OECD 2009 statistics on health spending, the United States spends the most ($956 per capita) for pharmaceutical and other medical goods—significantly higher than the $383.86 per capita average for 33 countries including the United States. In 2000, the United States spent an overall of $4,793 per capita and in 2009 spending increased to $7,960, per capita (Figure 10.1).

In efficiency, access and quality, data show that the U.S. standing decreased from fifth in 2004 to seventh in 2007. However, even if the U.S. overall health expenditure is the highest, the government does not allocate as much funding for healthcare insurance for the public. Therefore, the U.S. population achieved top ranking for private insurance, spending a staggering $2,599.10 per capita; Canada ranked second and spends only $509.30 (Figure 10.2).

OECD data for 2007 to 2009 compares healthcare expenditures and efficiency of various nations' healthcare systems. In total expenditures defined by Gross Domestic Product (GDP), the United States ranked seventh in comparison to six other developed countries: United Kingdom (ranked 1), Australia (ranked 2), the Netherlands (ranked 3), New Zealand

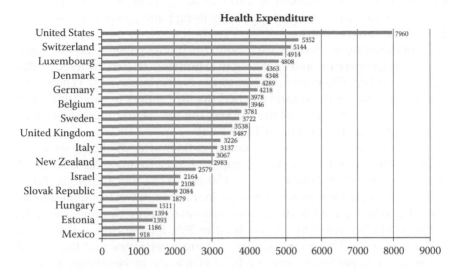

Figure 10.1 Health expenditure data for various countries in 2011. (Source: United Nations Organisation for Economic Cooperation and Development.)

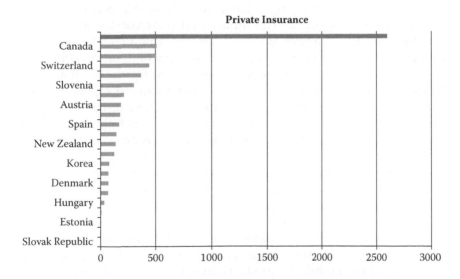

Figure 10.2 Private insurance expenditures of 11 countries excluding the United States in 2011. (Source: United Nations Organisation for Economic Cooperation and Development.)

(ranked 4), Germany (ranked 5), and Canada (ranked 6). These data also provide percentages and rankings describing patient experiences as outputs of the seven countries.

For example, when measuring how much time patients did *not* spend completing paperwork and engaging in arguments related to medical bills, the U.K. ranked 97% in comparison to the Netherlands' 68%. When measuring how efficiently records reached doctors' offices in time for patient appointments, the United States ranked the lowest and the Netherlands achieved the highest ranking (OECD 2011).

Hospital readmissions due to complications after discharge were frequent in the United States (18%) compared to Germany (9%), U.K. (10%), New Zealand (11%), Australia (11%), Canada (17%), and the Netherlands (17%). In use of information technology for clinical duties like maintaining records, ordering examinations, and dispensing prescriptions, Australia is the most effective at 91% in contrast to 26% for the United States and 14% for Canada. Based on the data, a pattern indicates that the United States has comparatively low levels of efficiency in healthcare systems (Tables 10.1 and 10.2; OECD 2011).

10.3   Productivity: General definitions and elements

Productivity is traditionally regarded as the relationship between the outputs generated by a system and the inputs needed to create the outputs (Sink 1991). Specifically, it is defined as the relationship of the amount produced by a system during a specific period and the quantity of resources consumed to produce the outputs over the same period (see Figure 10.3 flowchart). Typical inputs are in the form of labor (human resources), capital (physical and financial assets), energy, materials, and data. These resources are then transformed into outputs (goods and services). Different techniques exist to measure the costs associated with outputs and inputs (Stewart 1983).

In general, two forms of productivity are reported: partial and multifactor. Partial productivity is the ratio of gross or net output to one type of input among various inputs (Figure 10.3). Examples of the partial type are labor productivity and capital productivity. Total or multifactor productivity measures all outputs with all inputs. The U.S. Bureau of Labor Statistics (BLS) publishes productivity data for U.S. industries based on a number of sources of information about outputs and inputs used to calculate partial and multifactor productivities. Outputs are developed from Bureau of Census information as deflated values of production or physical quantities of production in certain industries. Inputs are specific to the type of partial productivity. Labor input is obtained by dividing the aggregate employee hours for each year by the base period aggregate. Capital

Table 10.1 Healthcare System Efficiency: Raw Percentage Scores for Seven Developed Countries

	Year	AUS	CAN	GER	NETH	NZ	UK	US
Total health expenditures (percent GDP)	2007	8.9	10.1	10.4	9.8	9	8.4	16
Percentage of national health expenditures for health administration and insurance	2007	2.6	3.6	5.3	5.2	7.4	3.4	7.1
No patient time spent for paperwork or disputes related to medical bills or health insurance	2007	90	88	86	68	87	97	76
Emergency room visits for conditions that could have been treated by regular doctor if available	2008	17	23	6	6	8	8	19
Records or test results did not reach doctor's office in time for appointment within past 2 years	2008	16	19	12	11	17	15	24
Sent for duplicate tests by different healthcare professionals, in past 2 years	2008	12	11	18	4	10	7	20
Hospitalized patients went to ER or rehospitalized for complication after discharge	2008	11	17	9	17	11	10	18
Practice utilizing high level clinical IT functions	2009	91	14	36	54	92	89	26

Source: United Nations Organisation for Economic Cooperation and Development Health Data. 2011. Netherlands data are estimated.

Table 10.2 Healthcare System Efficiency: Ranking Scores for Seven
Developed Countries

AUS	CAN	GER	NETH	NZ	UK	US
2	5	6	4	3	1	7
1	3	5	4	7	2	6
2	3	5	7	4	1	6
5	7	1.5	1.5	3.5	3.5	6
4	6	2	1	5	3	7
5	4	6	1	3	2	7
3.5	5.5	1	5.5	3.5	2	7
2	7	5	4	1	3	6

Source: United Nations Organisation for Economic Cooperation and Development:
Health Data. 2011. Netherlands data are estimated.

input is based on the flow of services derived from existing physical capital including equipment, structures, land, and inventories. Intermediate
purchases include real materials, services, fuels, and energy consumed
by an industry.

Productivity measures in healthcare settings use similar elements in
assessment. (Coleman et al. 2003, Phelps 2009). Inputs in healthcare studies, particularly those that use data envelopment analyses methods, consist of variables representing labor, capital assets, and/or other operational
expenses (Ashby and Altman 1992, Fraser et al. 2008, Freeman 2002, Mark
et al. 2009) that are readily known to nursing unit managers and used for
routine financial reporting. Labor, the predominant input, is measured in
healthcare settings by the number of hours of care per patient day by skill
mix level (for example, registered nurses, licensed practical nurses, and
unlicensed assistive personnel). Output variables may include patient satisfaction, medical errors, and patient falls (Freeman 2002, Hollingsworth
2008, Hollingsworth et al. 1999, Huttl et al. 2011).

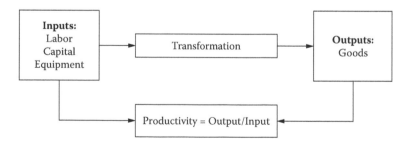

Figure 10.3 Flowchart of general framework for productivity assessment.

10.4 Factors affecting productivity

Several factors have been identified and studied for their impacts on the productivity of healthcare workers. We review the most important among these factors in this section.

10.4.1 Operational factors and politics of production

In healthcare systems, as in manufacturing, management faces a need to increase throughput (number of patients seen) while maintaining high levels of quality care and incentivizing productivity of staff through policy decisions and strategic use of resources. We briefly review these issues.

10.4.1.1 Managing quality versus quantity tradeoff

The tradeoff between quality and quantity is a long-standing issue that affects productivity and/or quality. Because achieving positive health outcomes for patients is critical in a healthcare system, the quality of care for every patient is as important as the number of patients seen by providers (Arrow et al 2012, Grieco and McDevitt 2012, Maudgalya et al. 2008). The key question is whether the quality of care is compromised when the number of patients treated by a system increases. Alternatively, will the quality of care increase when a healthcare provider treats fewer patients?

A balanced system with productive and quality outputs may be difficult to design and will require careful evaluation of inputs, throughputs, and outputs. For example, a study examining impacts of productivity and quality in an outpatient dialysis clinic found a tradeoff between quantity and quality (Grieco and McDevitt 2012).

A service system sustains itself through quality of service. It can be argued that efficient processing and transformation of inputs into outputs can determine quality. Some ways to achieve efficiency in transformation may be reducing patient waiting times, decreasing time spent with physicians, and delegating certain personnel to deliver specific services (Grieco and McDevitt 2012). However, if patients do not receive quality care, they are more likely to return often for treatment of the same condition. Every additional visit multiplies demands on a healthcare professional, effectively reducing the number of new patients seen. Conversely, patients may also be less likely to return to a facility because of poor care on a first visit, resulting in a revenue loss. The need for productive healthcare systems that focus on quality is critical and quality must be monitored closely and considered when designing incentive plans for providers.

10.4.1.2 Financial incentives and policies

In a healthcare system, cost reduction is imperative. However, increased quality of outcomes is needed as well. Balancing costs and quality is often a policy issue (Ashby and Altman 1992, Tambour 1997). Although healthcare outputs vary, the quality of care is a critical factor in delivery (Tambour 1997). The widely varying outputs including quality of care and patient safety demand careful design and consideration of financial policies. Effective compensation systems are crucial (Su et al. 2009, Tranmer et al. 2005).

Financial policies can lead to organizational changes and impact provider behavior. Ultimately, changes in financial matters can impact healthcare delivery. For example, if provider performance and compensation are determined on the basis of number of patients seen in a specific period, providers will face increased pressure to see more patients. Similarly, financial incentives may motivate healthcare organizations to adopt policies or practices that are not conducive to improving care quality or safety although they may increase productivity over the short term. To reduce fatigue, medical residents may have shortened work schedules but may be required to see the same number of patients. Thus, an increase in workload may be an unintended consequence.

Unfortunately, organizations, providers, and patients may not immediately feel the effects of these policies and financial incentive systems. The effects may be felt only as a cumulative problem over a long term or become apparent when an adverse event occurs. Flexible organization of care and motivating providers to strive for quality through a system of appropriate rewards may be the most effective measure for improving provider productivity (Baily et al. 1997).

At an individual level, fixed payments for services rendered to a patient, regardless of the time spent or costs incurred, can result in low quality care. Providers may decide to limit the time they spend on a patient service to allow them to treat more patients. A study of outpatient dialysis treatments (Grieco and McDevitt 2012) revealed that providers may intentionally or unintentionally reduce costs by decreasing quality of care.

Monetary incentives that reward physicians based on performance are more likely to increase their productivity. Studies show that individual production-based compensation may lead to increased productivity, whereas group-level financial incentives may not impact physician productivity significantly (Conrad et al. 2002). Bonuses also increase productivity even if they are not based on individual performance (Conrad et al. 2002). What is not clearly understood is whether individual production-based compensation adversely affects the quality of patient care.

10.4.1.3 *Staffing patterns and productivity*

A common perception is that increasing the number of physicians will increase productivity of an organization because productivity per physician is believed to be a constant number (Bloor et al. 2006). However, physician productivity depends on various factors such as their work-flows or work practice patterns, competence, or financial incentives. Adding staff may not lead to a direct increase in productivity. Increasing workforce size with new hires can help handle increased patient volumes temporarily, but organizational, individual, motivational, and financial problems can soon impact new workers and continue to challenge the system. Therefore, simply increasing staffing without alleviating interacting concerns may not improve productivity or quality (Bloor et al. 2006).

While increasing a workforce with new hires is one approach to maintaining productivity, increasing the number of individuals currently available for a task is another approach to productivity demands (Wilson et al. 2005). Studies have examined anesthesia team productivity by analyzing patient cases handled by teams and individuals. Team productivity was measured as mean monthly surgical anesthesia hours billed per attending anesthesiologist per clinical day. A supervisory ratio called concurrency (mean monthly number of cases supervised concurrently by attending anesthesiologists) was measured. Productivity was found to correlate positively with concurrency. Patient injury rate decreased with increased productivity, but the number of critical incidents (adverse events without adverse outcomes) increased (Posner and Freund 1999).

While sharing work responsibilities as team members may increase productivity, the additional demands for good communication, information flow, and mutual understanding of clinical situations among the team members may have led to the increased number of critical incidents. In essence, while the demands faced by individuals were shared, the cognitive work requirements were perhaps not as well distributed, possibly increasing productivity but affecting safety. It is important to consider staffing patterns based on a clear understanding of interactions resulting from shared responsibilities, patient demands, and the resulting quality of life of staff members (Brouwer et al. 2005).

10.4.2   Information technology and worker productivity

Information technology (IT) is a significant investment expected to improve healthcare quality dramatically. IT certainly displays the potential to improve the efficiency and effectiveness of healthcare delivery (Ko and Osei-Bryson 2004, Liederman et al. 2005, Petter et al. 2008). Computers can store large amounts of data and recall and process routine health information very quickly. They can also help generate reports from multiple data streams integrated into workflows. In general, routine healthcare tasks that are algorithmic and require little or no creative input can be handled by IT.

However, healthcare delivery is a service system in which people play a central role. Patients' disease conditions vary and in seeking healthcare, they interact with providers who have varying skill levels, training, knowledge, ability, and experience. The designer of a healthcare system must consider the reality that both IT and providers have capabilities and limitations. People can make decisions in unpredictable situations such as equipment breakdowns and effectively control the system until such situations are corrected. People are more flexible than IT and can easily adapt when systems change.

People also have limitations. They are very unpredictable and can create uncertainty; they have great difficulty in providing consistent quality, uniformity, reliability, and repeatability when they work (Mital and Pennathur 2004). They also have small working memory banks and may be unable to retrieve past information.

Technology is used to overcome the limitations of people. It can perform repetitive routine tasks consistently and aid in the performance of complex cognitive tasks. Health information technology also affords large data storage and processing capability and allows multitasking and integration of large datasets. However, technology has limitations as well. Routine use of technology may deteriorate the skills of providers.

In systems that use technology extensively, people supervise the system and their attention to and reliance on technology causes skill losses. Technology is generally inflexible and cannot easily adapt to changes (Mital and Pennathur 2004). Task artifact cycles (Carroll 1992) result from use of information technology. By the time healthcare workers learn to use a technological artifact to perform work, new technologies and designs impose additional learning burdens that impact their productivity. Health information technologies used by providers often require significant "workarounds" because of poor usability.

Lack of mode and situation awareness can be a critical problem arising from advanced health information technology. Automated systems often provide little or no feedback about their current status and future activities. As a result, workers are unaware of the inner workings of equipment and must perform significant cognitive work to infer current and future possible activities of the technology. Lack of mode awareness creates a gap between the mental models of humans and automated systems and operators may be unable to anticipate system behavior. This gap creates "automation surprises" that must be handled (Sarter and Woods 1997). A poor mental model match, low system observability, and fast-changing situations all contribute to automation surprises along with reduced productivity and effectiveness (Sarter and Woods 1997).

Technology, as part of patient interaction, should be designed, as an enabling tool to support the task demands of providers caring for patients. It should not be a burden on providers. Mere investments in technology cannot improve healthcare worker productivity. Instead, designs that enhance effective use of the technology are more likely to increase the effectiveness of care and improve productivity. Devaraj and Kohli (2003) found that technology use based on well-designed systems was associated positively with hospital revenue and quality. Providers using film-less technologies for imaging required less examination time than users of film-screen radiography (Reiner and Siegel 2002). Carefully planned changes in technology that utilize functionality to support the tasks of end users seemed to increase productivity and quality of life.

However, fear of new technology, and resistance to adapting to and using IT must be anticipated and handled. Even if technology will improve performance, it may introduce problems of autonomy and knowledge control (Drucker 1999, Ramírez and Nembhard 2004). Some providers may perceive new IT to have negative effects on their performance and autonomy over the long term (Succi and Walter 1999).

Another issue is the impact of financial incentives on using IT for providing care (Lodge 1991). Providing care via electronic communications may create an issue related to lack of reimbursement for sending patient messages (Liederman et al. 2005). If providers are not paid for time they spend communicating with patients electronically and if the system does not store their communications, the providers are unlikely to use the technology to communicate with patients.

If technologies are designed so providers understand that electronic communication may increase their productivity and help them better prioritize their time, they may change their practices and use the technology. In a study investigating the use of an electronic message system for patients, the productivity of physicians communicating with their patients using a web messaging system was compared to a group of physicians not using such systems. Physicians who used the electronic system were 10% more productive than those who did not. The productivity increase led to 2.54 more patient visits (Liederman et al. 2005).

Because web messaging is asynchronous and ubiquitous, the physicians could communicate with patients from their homes at convenient times rather than from their workplaces. The flexibility freed up more time for patient visits during work hours. In addition, patients who completed their histories electronically spent less time in physicians' offices. Physicians who understood the potential benefits of web messaging were able to adjust their workflows and work practices to increase their productivity (Liederman et al. 2005).

Although the study raises the question of whether the quality of care will be different based only on use of an electronic communication medium in combination with an office visit, the number of physicians now using this mode of patient care lends credence to the idea that IT in healthcare has significantly impacted provider work practices and shows the potential to increase productive time spent with patients significantly.

The IT productivity paradox (Ko and Osei-Bryson 2004) has long been debated and is an important issue that deserves more attention. The complex interactions of IT, healthcare providers, and patients and IT's impacts on productivity and care quality require further research.

10.4.3   Quality of working life

Unbalanced staffing patterns and work distribution can affect the quality of provider work life significantly and cause decreased productivity. (Ashby and Altman 1992, Brouwer et al. 2005). For example, acute care nurses are subject to very high workloads, and hence have no time to improve their performance (Brooks and Anderson 2004).

Large numbers of physicians or nurses (input resources) available to perform the work can indicate higher productivity levels. However, mere presence does not directly indicate or predict productivity. For example, "presenteeism" is a major nursing issue that arises when employees are present but cannot perform productively due to illness or a condition such as depression or stress (Pilette 2005). It is important to consider that the health status of providers affects their abilities to be productive. Worker health may be affected by multiple factors including work schedules, domestic issues, sociodemographics, and abilities to cope with stresses and demands (Curtin 1995, Eisenberg et al. 2001, Gates et al. 2011, Kc and Terwiesch 2009).

A study by Fischer et al. (2006) examining living and work conditions of nurses in a public hospital in Brazil used a questionnaire to collect data on incivility at work, work demands, control, support, fatigue, sleep, physical health, and abilities. Significant effects on ability to work arose from sociodemographic issues such as the need to sustain income and provide for children, working conditions, verbal abuse from patients and supervisors, and environmental deterrents and health outcomes (high body mass index [BMI], obesity, sleeplessness, and fatigue). Additionally, stressful work conditions and demographic constraints were found to impact the ability of nurses to work productively (Fischer et al. 2006). These findings call for intervention measures, for example, redesigning processes to better fit the needs of the nurses, and setting policies that promote safe and civil working conditions (Gates et al. 2011).

Work schedules play an important role in safe working conditions and improving the quality of working lives of providers. A study by Poissonnet and Véron (2000) investigating the effects of work schedules on the quality of working life and job performance found that no single system was more effective than another but extended workdays (9- to 12-hour shifts) are not recommended.

Other studies revealed that work schedules impact provider performance and also patient safety. For example, nurses working longer shifts (more than 12.5 hours) are more likely to be less vigilant, increasing the likelihood of errors or occupational injuries. Similarly, physicians who work more than 24 hours are more likely to sustain injuries or make serious errors (Lockley et al. 2007).

Clearly, extended work hours can significantly cause fatigue and impact performance and safety (Debra et al. 1997, Poissonnet and Véron 2000, Tanabe and Nishihara 2004). While safe work hour limits are essential, we need more evidence to understand whether limits foster effective work because hour limits may require providers to handle the same workloads in shorter times (Lockley et al. 2007). Perhaps, IT systems that do not suffer fatigue can be utilized to support providers by performing routine tasks to improve provider work life and productivity.

10.4.4   Personal characteristics and productivity

Individual characteristics such as experience and skill may determine how providers adapt to the changing demands of healthcare systems and how they balance productivity, quality, and safety in care delivery (Landon et al. 2001, Letvak and Buck 2008). Health providers, like other professionals, must be competent in their areas of specialty. Competence implies having the necessary skills, knowledge, and characteristics to perform work adequately and is the primary determinant of productivity. Although a provider may have essential competence, it may not be used to its full extent due to external pressures or individual attributes (Kak et al. 2001).

While efficiency is often cited as a major factor in productivity, effectiveness in job performance is a more important issue affecting productivity. Comparing the levels of provider competence and actual performance using such competence can reveal problem areas such as a lack of organizational support for professional development and training and insufficient resources to perform effectively. Higher competence levels alone do not always ensure better performance (Kak et al. 2001).

Other characteristics of a healthcare provider that may affect productivity include motivation, social conditions outside work, and feedback and reward for performance (Kak et al. 2001). Although a provider may be technically competent to perform work, he or she may not be motivated by financial incentives. A provider may be reluctant to change or adapt to new ways of working (e.g., use IT) if he or she perceives reduced incentives or thinks patient safety will be compromised.

Studies on the impacts of individual characteristics on performance portray mixed results. For example, in a study examining resident performance, personal characteristics showed few associations with performance compared to other factors such as healthcare infrastructure, medical education infrastructure, and process measures such as knowledge, skills, attitudes, habits, and health outcomes (Michael et al. 2005).

Because of the high degree of interaction between providers' work practices and the resources used to perform work, it is important to examine the interaction between provider characteristics and workflow to improve productivity outcomes.

Studies report that the workflow and practice patterns of individual physicians are more significantly associated with their productivity than with clinic (organizational) or patient characteristics. For example, a decentralized appointment system seemed to increase physician productivity by reducing no-show visits; providing education to patients during vaccination visits seemed to decrease physician productivity. The authors opine that delegating education responsibilities to nursing staff and taking responsibility for direct care alone may increase physician productivity (Smith et al. 1995).

When financial incentives seemed to increase physician productivity, their practices were influenced accordingly and further increased productivity (Smith et al. 1995). While time spent with patients depends on several factors including disease complexity, medications, an emergency room visit within the past 6 months, and other items, *how* providers spent their time and what actions and behaviors they exhibited during patient encounters seemed to affect productivity more than amounts of time spent with patients. New policies, personnel changes, and technologies may unintentionally produce positive and negative impacts on the work practices of providers.

10.5   Summary

Healthcare delivery is an emerging service industry involving investments of billions of dollars for capital infrastructure, labor, and other resources. Ineffective healthcare delivery practices directly affect patient well-being. This fact exerts increased pressure on healthcare workers to increase their productivity and effectiveness.

Our review indicates that staff productivity and effectiveness are functions of work design. Optimizing healthcare delivery requires strategic goal setting for improved patient outcomes, effective work design, and efficient management of skilled human capital. Knowledge workers in a healthcare setting possess specialized skills and expertise. They must innovate continuously in demanding environments.

Organizational support structures that permit and promote innovative healthcare delivery are critical if healthcare workers are to be productive and effective. Performance measurement and evaluation paradigms that promote and reward quality along with quantity of care, provide well-designed IT tools that simplify and aid the complex work of healthcare, and implement operational policies that emphasize a high quality of work life for staff can improve productivity and foster effective healthcare delivery.

References

Arrow, K., Bilir, K., Brownlee, S. et al. 2012. Valuing Health Care: Improving Productivity and Quality. Report by Kaufmann Task Force on Cost-Effective Health Care Innovation. http://papers.ssrn.com/sol3/papers.cfm?abstract_id=2042644. Accessed April 30, 2013.

Ashby, J.L., Jr. and Altman, S.H. 1992. The trend in hospital output and labor productivity, 1980–1989. *Inquiry*, 29, 80.

Baily, M.N., Garber, A.M., Berndt, E.R. et al. 1997. Health care productivity. *Brookings Papers on Economic Activity. Microeconomics*, 18, 143–215.

Bloor, K., Hendry, V. and Maynard, A. 2006. Do we need more doctors? *JRSM*, 99, 281–287.

Brooks, B.A. and Anderson, M.A. 2004. Nursing work life in acute care. *Journal of Nursing Care Quality*, 19, 269.

Brouwer, W.B.F., Meerding, W.J., Lamers, L.M. et al. 2005. The relationship between productivity and health-related QOL: an exploration. *PharmacoEconomics,23*, 209–218.

Carroll, J.M. 1992. Getting around the task artifact cycle: how to make claims and design by scenario. *ACM Transactions on Information Systems*, 10, 181–212.

Coleman, D.L., Moran, E., Serfilippi, D. et al. 2003. Measuring physicians' productivity in a Veterans Affairs medical center. *Academic Medicine*, 78, 682.

Conrad, D.A., Sales, A., Liang, S.Y. et al. 2002. The impact of financial incentives on physician productivity in medical groups. *Health Services Research*, 37, 885–906. doi: 10.1034/j.1600-0560.2002.57.x

Curtin, L.L. 1995. Nursing productivity from data to definition. *Nursing Management*, 26, 25.

Debra, R., Gregory, M.R., Mallis, K.B. et al. 1997. The cost of poor sleep: workplace productivity loss and associated costs. *Journal of Occupational and Environmental Medicine*, 52, 91. doi: 10.1097/JOM.0b013e3181c78c30

Devaraj, S. and Kohli, R. 2003. Performance impacts of information technology: is actual usage the missing link? *Management Science*, 49, 273–289.

Drucker, P.F. 1999. Knowledge worker productivity. *California Management Review*, 41, 79–94.

Eisenberg, J.M., Bowman, C.C., and Foster, N.E. 2001. Does a healthy healthcare workplace produce higher quality care? *Joint Commission Journal on Quality and Patient Safety*, 27, 444–457.

Fischer, F.M., Borges, F.N., Rotenberg, L. et al. 2006. Work ability of healthcare shift workers: what matters? *Chronobiology International*, 23, 1165–1179. http://dx.doi.org.proxy.lib.uiowa.edu/10.1080/07420520601065083

Fraser, I., Encinosa, W., and Gllied, S. 2008. Improving efficiency and value in health care: introduction. *Health Services Research*, 43, 1781–1786.

Freeman, T. 2002. Using performance indicators to improve health care quality in the public sector: a review of the literature. *Health Services Management Research*, 15, 126–137. doi: 10.1258/0951484021912897

Gates, D.M., Gillespie, G.L., and Succop, P. 2011. Violence against nurses and its impact on stress and productivity. *Nursing Economics*, 29, 59–66.

Grieco, P.L.E. and McDevitt, R.C. 2012. *Productivity and Quality in Health Care: Evidence from the Dialysis Industry.* University Park, PA Pennsylvania State University and University of Rochester.

Hollingsworth, B. 2008. The measurement of efficiency and productivity of health care delivery. *Health Economics,* 17, 1107–1128.

Hollingsworth, B., Dawson, P.J., and Maniadakis, N. 1999. Efficiency measurement of health care: a review of nonparametric methods and applications. *Health Care Management Science,* 2, 161–172.

Huttl, A., Mas, M., Nagy, A. et al. 2011. Measuring the productivity of the health-care sector: theory and implementation. Report from the Indicators for Evaluating International Performance in Service Sectors (INDICSER) Group.

Kak, N., Burkhalter, B., and Cooper, M.A. 2001. Measuring the competence of healthcare providers. *Operations Research Issue Papers,* 2, 1.

Kc, D.S. and Terwiesch, C. 2009. Impact of workload on service time and patient safety: an econometric analysis of hospital operations. *Management Science,* 55, 1486–1498.

Ko, M. and Osei-Bryson, K.M. 2004. The productivity impact of information technology in the healthcare industry: an empirical study using a regression spline-based approach. *Information and Software Technology,* 46, 65–73.

Landon, B.E., Reschovsky, J., Reed, M. et al. 2001. Personal, organizational, and market level influences on physicians' practice patterns: results of a national survey of primary care physicians. *Medical Care,* 39, 889.

Letvak, S. and Buck, R. 2008. Factors influencing work productivity and intent to stay in nursing. *Nursing Economics,* 26, 159.

Liederman, E.M., Lee, J.C., Baquero, V.H. et al. 2005. The impact of patient–physician web messaging on provider productivity. *Journal of Healthcare Information Management,* 19, 81–86.

Lockley, S.W., Barger, L.K., Ayas, N.T. et al. 2007. Effects of healthcare provider work hours and sleep deprivation on safety and performance. *Joint Commission Journal on Quality and Patient Safety,* 33 (Suppl. 1), 7–18.

Lodge, D. 1991. Productivity, efficiency, and effectiveness in the management of healthcare technology: an incentive pay proposal. *Journal of Clinical Engineering,* 16, 29.

Mark, B.A., Jones, C.B., Lindley, L. et al. 2009. An examination of technical efficiency, quality and patient safety in acute care nursing units. *Policy, Politics, and Nursing Practice,* 10, 180–186.

Maudgalya, T., Genaidy, A., and Shell, R. 2008. Productivity, quality, costs, and safety: a sustained approach to competitive advantage. *Human Factors Management,* 18, 152–179. doi: 10.1002/hfm.20106

Michael, M., Maya, S., Malathi, W. et al. 2005. Factors affecting resident performance: development of a theoretical model and a focused literature review. *Academic Medicine,* 80, 376.

Mital, A. and Pennathur, A. 2004. Advanced technologies and humans in manufacturing workplaces: an interdependent relationship. *International Journal of Industrial Ergonomics,* 33, 295–313.

OECD. 2011. Health Data. http://stats.oecd.org/Index.aspx?DataSetCode=SHA

Petter, S., DeLone, W., and McLean, E. 2008. Measuring information systems success: models, dimensions, measures, and interrelationships. *European Journal of Information Systems,* 17, 236. doi: doi:10.1057/ejis.2008.15

Phelps, M. 2009. Total public service output and productivity. *Economic and Labour Market Review,* 3, 45–55.

Pilette, P. 2005. Presenteeism in nursing: a clear and present danger to productivity. *Journal of Nursing Administration,* 35, 300.

Poissonnet, C.M. and Véron, M. 2000. Health effects of work schedules in healthcare professions. *Journal of Clinical Nursing,* 9, 13–23.

Posner, K.L. and Freund, P.R. 1999. Trends in quality of anesthesia care associated with changing staffing patterns, productivity, and concurrency of case supervision in a teaching hospital. *Anesthesiology,* 91, 839.

Ramírez, Y.W. and Nembhard, D.A. 2004. Measuring knowledge worker productivity: a taxonomy. *Journal of Intellectual Capital,* 5, 602–628. doi: 10.1108/14691930410567040

Reiner, B.I. and Siegel, E.L. 2002. Technologists productivity when using PACS: comparison of film-based versus filmless radiography. *American Journal of Roentgenology,* 179, 33–37.

Sarter, N.B. and Woods, D.D. 1997. Team play with a powerful and independent agent: operational experiences and automation surprises on the airbus A-320. *Human Factors,* 39, 553.

Sink, D.S. 1991. The role of measurement in achieving world class quality and productivity management. *Industrial Engineering,* 23(6), 23–30.

Smith, D.M., Martin, D.K., Langefeld, C.D. et al. 1995. Primary care physician productivity. *Journal of General and Internal Medicine,* 10, 495–503. doi: 10.1007/BF02602400

Stewart, W. 1983. Multiple input productivity measurement of production systems. *International Journal of Production Research,* 21, 745–753.

Su, S., Lai, M.C., and Huang, H.C. 2009. Healthcare industry value creation and productivity measurement in an emerging economy. *Service Industries Journal,* 29, 963–975.

Succi, M.J. and Walter, Z.D. 1999. Theory of user acceptance of information technologies: an examination of health care professionals. In *Proceedings of 32nd Annual Hawaii International Conference on System Sciences.* Maui, Hawaii, January 5–8. http://ieeexplore.ieee.org/xpl/login.jsp?tp=&arnumber=773013&url=http%3A%2F%2Fieeexplore.ieee.org%2Fxpls%2Fabs_all.jsp%3Farnumber%3D773013

Tambour, M. 1997. The impact of health care policy initiatives on productivity. *Health Economics,* 6, 57–70.

Tanabe, S. and Nishihara, N. 2004. Productivity and fatigue. *Indoor Air,* 14, 126–133.

Tranmer, J.E., Guerriere, D.N., Ungar, W.J. et al. 2005. Valuing patient and caregiver time: a review of the literature. *Pharmacoeconomics,* 23, 449–459.

U.S. Bureau of Labor Statistics. http://bls.gov/mfp/

Wilson, K.A., Burke, C.S., Priest, H.A. et al. 2005. Promoting healthcare safety through training high reliability teams. *Quality and Safety in Health Care,* 14, 303–309.

Index